WINDMILLS OF SOMERSET

AND THE MEN WHO WORKED THEM

WORLE MILL

A retouched photograph of 1867, probably the earliest surviving Somerset windmill photograph. The almost cylindrical tower, gable-shaped cap and broad sails were typical in the county.

WINDMILLS OF SOMERSET

AND THE MEN WHO WORKED THEM

---

by ALFRED J. COULTHARD
and MARTIN WATTS

WITH A FOREWORD BY
Rex Wailes

THE RESEARCH PUBLISHING CO.
52 Lincoln's Inn Fields · London

Also by Martin Watts

VICTORIAN AND EDWARDIAN WINDMILLS AND WATERMILLS
(with J. Kenneth Major)

ISBN 0 7050 0060 5
Printed in Gt. Britain for The Research Publishing Co.
(Fudge & Co., Ltd.), London.

CONTENTS

FOREWORD

It was in 1931 that I first visited Ashton Mill in Somerset and talked to Mr Stevens, the last owner to operate it. Twenty-nine years later, when it had been splendidly restored at the charge of the late Mr Cyril Clarke, I was present at the opening which he himself performed, and presented him with a windmill certificate of the Society for the Protection of Ancient Buildings. In 1966 Mr Clarke presented the mill to the City Museum, Bristol, and in the following year I had the pleasure of leading their first Conducted Walk to the mill.

For many years I have known Captain Coulthard and Mr Watts as leading authorities on the windmills of Somerset and it is fortunate indeed that they have recorded their deep knowledge of their history, constitution and recollections of the men who worked them. Their work on this specialised aspect of local history should be widely welcomed by all who are interested both in Somerset and in the subject of windmills, for it enables one to realise the continuity of windmilling over the centuries and the difficulties of those who had to work the mills. How from being tenants they gradually became owners and later landlords of the mills, at first grinding others' grain for payment in kind, then for money and finally becoming corn merchants and, in some cases, bakers, before the eventual decline for reasons so graphically described by one of their company in this book.

By 1931 no windmills remained in the country, either wholly or partly complete, with a resemblance to those of Somerset. Perhaps those of Cornwall or along the Welsh coast of the Bristol Channel may have once resembled them, but they are represented only by the shells of their towers and so far there does not appear to be on record any first-hand evidence of their internal economy.

The Somerset mills were stone built and had caps, some thatched, which were turned by hand chain to face the cloth-spread sails into the eye of the wind, while many drove only a single pair of stones. Though not so advanced in design, the nearest complete comparable examples known to me were to be seen in parts of Brittany.

Fortunately, in addition to Ashton Mill, Stembridge Mill, High Ham, is also preserved and belongs to the National Trust. They cannot be appreciated fully without the background provided by the valuable illustrated records collected by the authors and set out for our pleasure in this book.

REX WAILES
O.B.E., F.S.A., F.I.MECH.E., F.F.S.A.

ACKNOWLEDGEMENTS

The authors wish to acknowledge with grateful thanks the help of the many individuals and institutions who have made this collection of information on Somerset windmills possible, in particular Dr R. W. Dunning of the *Victoria County History for Somerset*; the Science Museum for information from the collection of the late H. E. S. Simmons; Godfrey Shove of Hutton for his work on the Tithe Awards and other archival work and for searching out many valuable photographs; the staff of the Somerset Record Office; Rex Wailes for his interest and advice and Kenneth Rogers of the Wiltshire Record Office for bringing attention to the Longleat Papers deposited there. Also the Editors of *Millnotes* and *Notes & Queries for Somerset & Dorset* for permission to include extracts here.

The illustrations are from the authors' own collections, with the exception of the following which are by courtesy of :

Dr. R. W. Dunning (Pibsbury Mill); Mrs. G. M. Evans (Walton Mill, right); Frank Gregory (Vale Mill, spur wheel); Mrs. E. Norman (Portishead, top); Godfrey Shove (Bishop's Lydeard, Ditcheat, North Cadbury and Heath House Mill); SPAB (Watchfield Mill, right, and Willtown Mill, left); Somerset Record Office (Week, Curry Rivel—photograph by Godfrey Shove); Taunton Museum (Moorlinch); Rex Wailes (Ashton Mill, sack hoist, Vale Mill, brake wheel and secondary drive); Woodspring Central Library (Frontispiece, Ashton Mill, pre 1897, T. Quick at Vale Mill and Worle Mill).

Finally this book could not have been compiled without the courtesy and help given by many Somerset millers and their families and the owners of the mills and sites, past and present.

AUTHORS' NOTE

It should be appreciated that the county of Somerset referred to in this book is the historic county before the boundary changes of 1974.

INTRODUCTION

This book is the result of two independent pieces of research, separated by a space of nearly fifty years but joined by kindred spirits and common aims. The first is largely a personal record made, without any intention of publication, by Captain Coulthard between about 1920 and 1934 of windmills and windmill sites visited and examined, and of millers chatted to whenever the opportunity arose. Sometimes these chats were repeated at intervals over several years. A few additions were made to these notes after 1934 on subsequent occasional visits to the county but the notes were compiled at a time when, but from the millers themselves, there was little encouragement to acquire information about windmills and the total number of those actively interested all over England could probably be counted on both hands and were generally considered a queer cult. Between 1930 and 1950, for example, many appeals were made for assistance in the preservation of Ashton Mill and other Somerset windmills without ever kindling any positive action.

The second study began in about 1969 when a new search for information on Somerset windmills proved it to be a rather neglected subject even in an improving climate and was considerably impeded by the fact that the last windmill to work in the county had ceased milling in 1927 and all the men who had worked directly with the mills were gone. Research was heightened a little, however, by the accessibility of an increasing amount of historical information, from which often minimal facts were eagerly gleaned.

The result of these two independent researches is amalgamated in this account, in part intending to depict the scene at the end of the long tradition of Somerset windmills some fifty years ago from the people who were part of it, and in part recording the facts and remains of the windmills from field work and documentary sources. As a whole it is intended to arouse interest and promote some knowledge of this previously little-regarded aspect of Somerset's history, and both authors are pleased that the publisher has enabled them to come together and record in joint form their researches. Let Captain Coulthard now " set the scene "...

1 SETTING THE SCENE

It is not enough to approach the subject of Somerset windmills as merely another inventory of an as yet relatively untouched field of industrial archaeology, and, however unsuited I may be to describe atmosphere, I must, I think, make some attempt to portray an accurate perspective.

On the Scottish border, both sides of it, memories exist in an almost uncanny form of happenings of centuries ago; one is shown a standing stone or a grass-grown track : ' The reivers used to meet there ' or ' That is the way they used to come '. When? ' Oh, that would be before my grandfather's time '. Somerset was not like that. Memories were much shorter and, on the whole, limited to trees on which Monmouth's men were hanged and barren hilltops on which corn was grown in the Napoleonic Wars, although a farmer's wife at Vaggs Hill, near Tellisford Mill, did once show me a track ' where Cromwell brought the cannon down ', which I believe to be correct. Again, some parts of Somerset have not been populated that length of time. There was an influx when the moors were drained, between say 1770 and 1830, and a lot of new land was made available for farming. This may not have affected a great proportion of the county considered overall, but it did affect much of the land between the Mendips and the River Parrett, with which we are particularly concerned. I mention this because, as I say, on the whole memories did not as a rule go back beyond ' grandfather's day ', whereas on the Border they would have done. This, then, is one feature.

Another feature which I must touch upon to forestall possible criticism of my recordings of chat and conversation, is that at least as often as not something like the following would take place at some point, perhaps the first point ! ' How about a drop of cider? '. 'How do you like it, sweet or dry? '. ' Come along into the cider house and see which you like best '. I suppose it is necessary now to describe a Somerset cider house which was a long, low, narrow, dark building, with casks each side of a narrow path placed on low platforms. The miller-cum-farmer would stop at a cask, tap it with his hand and say ' In an old brandy cask that lot. Care to try it? '. A drinking vessel —or something one could drink out of in the dark—was filled and you agreed that this was really good, but your host had passed on to another cask. 'An old wine cask this. Perhaps you would like this better? '. This could go on indefinitely and getting back to the house from the cider house was quite a problem because, of course, of the sun in your eyes after the dark cellar.

However, back to windmills and millers. Mr Duckett, the last miller at Westfield Mill, told me in about 1930 : ' There was no spring carts in either of these places (Wedmore, Stoughton, Ashton, Mark and Heath House) and if people hauled any wheat or beans to any of these mills, they could only drive at walking pace as there were only putts to get about with. A good many rode on horse-back and carried what they could. They had a thing each side of the horse's back; the person got in one, one side, and whatever they had to carry was put in the other side with the fastening across the back of the horse. I have heard my mother say that she used to go from East Brent to Axbridge market with two baskets of butter every Saturday in this way. Axbridge is a very old place and there was a butter and sheep market there every Saturday afternoon. I have also heard my mother say that they used to court this way. When they wanted to go far, the man sat one side of the horse and the young woman the other. This would be about eighty-five years ago. The first mowing machine in Wedmore my father (Michael Duckett) had about the year 1873 and people came from miles around to see it work. Of course they had to walk or ride on horse-back '.

He also remarked : 'Windmills were used to grind all the wheat into flour years ago, and also all the beans into meal for stock to eat as there was no maize in this county. Of course the flour was made into bread, as every farmer's wife baked her own bread. There were no bakers' carts going around, but there was a baker at Blackford and one at Wedmore two miles away, and if you wanted a loaf of bread any time you could have it by fetching it. A lot of the bread had been baked ten days ! ' As an indication of the amount of business done, Mr William Spearing, miller at Isle Brewers, told me : 'In those days the millers used to bake and sell brown bread. Aunt Mary was the eldest daughter and she used to tell how she saved enough halfpennies to buy a tea service whilst selling bread '.

Another miller told me that on rare occasions it became necessary to make the trip to Bristol. This trip would have to wait until there were others to make the journey, all on horseback, because one man on his own would certainly be robbed of his purse somewhere along the road. There was a notorious public house on their route where the foot-pads or highwaymen, as he called them, used to hang about. Once past that they could relax a bit.

Mr Edwin Edwards, who was still alive in 1925 and who as a boy had worked on my maternal grandfather's farm at Edingworth, often told me how he remembered the men coming back to the farm from the Crimean War. He also remembered one of the men on the farm being transported for sheep-stealing. When he knew I was interested in windmills, he told me that as a boy he often used to go up on Brent Knoll and sometimes see no less than seven windmills at work. He was not at all certain which mills they all were, but he was quite certain about the number. All he could recollect was that 'there were three at Allerton '. These three would presumably have been Weare, Ashton and Stoughton and the others probably Blew's, Watchfield, Moorlinch and either Woolavington or Heath House. I asked Tom Petheram about this and he rather doubted Weare Mill, but agreed about the others. I also asked him about highwaymen and he said 'Father told me that in those days for their own safety the butchers and millers etc., had to form a convoy when going to and returning from towns to the various villages, as the road was long and lonely and plenty of Gents waiting for a lonely man's gold'. Talking of Middle Stoughton, where John Burrows of Stoughton Mill lived, Tom Petheram said, ' I knew his neighbours too, Ben and Jim Higgs, two of the real old Somerset type, kittle (?) smock, tight breeches, ash stick and red handkerchief around the neck. They lived in what must have been a large farmhouse, they had been farmers and butchers, used to kill calves and pigs etc., and take the meat to Bristol by road. They also made a lot of cider and took that to Bristol too. I remember the shoe-maker there too, he used to start out about 4 a.m. to walk into Bristol to buy his leather, to be sent out later, walk home to dinner and do his usual day's work after '.

Regarding Mr Edwards' ' seven windmills ', Mr Stevens, the last miller at Ashton Mill, agreed with my suggested listing saying that Moorlinch Mill used to catch the sun with its white-washed tower and black sails often when things much nearer could not be seen. When he first worked at Ashton Mill he would see five mills working if the wind was suitable for every one. ' These would be Moorlinch, Heath House, Watchfield, Woolavington and one other'. I suggested Weare Mill, but he did not think that still worked in his day. When I asked Tom Petheram he said ' he forgot his own mill '.

2 HISTORICAL BACKGROUND

Although much work has still to be done to compile a full record of the windmills which were built and worked in Somerset from the early 13th century to the late 19th century, and this book is concerned mainly with the windmills and millers at the end of this long tradition, it is obviously worthwhile to outline the history of windmilling in the county to give a true perspective.

A valuable source of early information, the Longleat Manuscripts, notably dealing with the mills connected with Glastonbury Abbey, is being studied by Mr Ian Keil,[1*] and the present editor of the *Victoria County History for Somerset,* Dr R. W. Dunning, has contributed much useful information on mills in the parishes so far covered by the History.[2] It is probable that more records and information will come to light with the increasing interest in local history and the history of technology and until all the parishes in the county have been exhaustively covered it is not possible to know the full story, but from information already gathered on windmills in Somerset it is possible to outline their development.

By the time the windmill made its appearance in Western Europe towards the end of the 12th century, grain mills, powered by man, animals or water, were already commonplace in England. The Domesday Survey records many in Somerset, although none of these were windmills. The earliest reference to a windmill in the county is contained in the Montacute Chartulary and refers to one Robert de Vallibus who granted to the church of Montacute ' together with his body, his wind-mill of Sevenamtune, with the grinding of the whole of the same manor, and with six acres of land, close to the said wind-mill, from his own demesne'.[3] Sevenamtune is the present day Seavington St Michael and although this reference is only put in the early 13th century in its original context Dr Dunning has dated it more precisely to circa 1212, just over a quarter of a century after the first authentic reference to a windmill in England. In 1222 the tithes of two windmills in Woolavington were given to the monks of Goldcliff in Monmouthshire by Philip de Columbers the third, ' for the soul of Philip his first-born ' and it was also granted that ' their servants may come to the mills and grind their corn " after the first corn they may have found in the hopper " '.[4] From the middle of the 13th century, references to windmills become more common, there being windmills at Glastonbury in 1244 and Street in 1255.[5]

One of the most fascinating accounts of a mediaeval windmill comes from

*Notes appear under the heading "Notes and References ", page 71.

the Longleat Manuscripts and refers to the building of a post mill at Walton in 1342-3.[6] This mill was built, probably as a capital improvement, by Glastonbury Abbey; it cost a total of £11.12s.11d to construct and in 1358 the annual rent stood at £3. Although the costs and, in some instances, exact quantities of the materials used are given in the accounts, it is not possible to reconstruct an accurate picture of this mill. If it may be assumed that mediaeval post mills were fairly consistent in design and appearance throughout England, then this early mill at Walton was probably similar to the post mills depicted in mediaeval illustrations.

No early windmills have survived in the county and there is no full description of a Somerset post mill, but several sites and mill mounds have been excavated and the remains of the sub-structure of some early windmills uncovered. At a site opposite Sydenham Manor, to the north east of Bridgwater, a mound was excavated in the path of a proposed M5 access road in 1971 and probably the most complete set of timbers yet found was removed. The main post was missing, but remains of both cross trees and the quarter-bars which supported the post were found buried in this mound of late mediaeval make-up. Similar, lesser, remains have been uncovered at other sites in the county and these are recorded in the list of sites.

There are five notable depictions of post mills in Somerset and, while they must be regarded as stylised, they are a valuable part of the county's windmill history. Two wall paintings, originally of 15th-century date, exist in the parish churches at Ditcheat and Henstridge. That at Ditcheat, on the north wall of the nave, shows St Christopher with a small post mill to his right side, and has been much restored, whilst at Henstridge is a copy of a destroyed painting, again of St Christopher, with a post mill, a laden packhorse, a dog and a man carrying a sack. Perhaps the proximity of windmills to the main roads of the county and their value as landmarks prompted this association with the saint of travellers. At Bishop's Lydeard and North Curry fine bench end carvings survive, that in the former church showing a post mill in front elevation, with three birds, a sack of corn, a man and a saddled horse, and is of 16th-century date. At North Curry the carving shows a post mill in side elevation with a millstone by it, and is dated to 1538. Another bench end carving at East Lyng church shows a miller riding a horse and carrying a sack, a common late mediaeval illustration, but shows no mill. On the metope frieze between 13 and 14 The Circus, in Bath, appears a simple carving in Bath stone of a post mill. The Circus was built between 1754-8, and other carvings on the frieze include two pairs of millstones and an overshot waterwheel.

All these illustrations are notable in that the sub-structures of the mills portrayed appear to be buried in mounds or in the ground, and this is consistent with archaeological finds and documentary evidence that some post mills were built with their trestles buried or ' sunk ' in the ground, presumably to give them increased stability in exposed locations. This type of windmill is known as the sunk post mill and examples have been found, through excavation, throughout England.

Exactly how many windmills were at work in Somerset at any one time from the later middle ages until the 18th century is rather a matter for conjecture, for while many sites have been occupied by windmills for two hundred years, and some for longer, it is not until the mid 17th century that maps appear which show windmills together at a specific point in time. Ricardus Newcourt's map of ' Sedgemoor, with adjacent places ' of c.1660, marks some 17 windmills, mostly on the higher ground surrounding the Moor. John Ogilby's Road Maps of c.1675 are also a useful source, marking windmills as landmarks, sometimes by name, near the main travelling routes. The map of the road from London to Barnstaple, for example, shows 10 windmills close to the ridgeway across the Polden Hills within a distance of eight miles. There appears to have been a decline in the number of mills standing by the mid 18th century, for Emanuel Bowen's County Map of c.1760 shows only five mills on the Polden Ridge, and only 16 in the whole county, although this is not evidence that there were not more mills at that time. It does seem, however, that many of the windmills whose existence is perpetuated in official place names such as Windmill Hill, Windmill Farm and Mill Batch, had gone out of use by this time, and many were never replaced.

By the early 15th century the tower mill had become established in Europe, and offered in general terms a more stable structure than the timber post mill. The first records of tower mills in Somerset date from the early 18th century but while elsewhere the tower form was used to increase storage facilities and actual output, the Somerset tower mill was generally of modest stature and ouptut, although efficient, and probably was the product of small acreages and prevailing wind currents. The decline of windmilling in the county, and perhaps in the south west of England generally, was earlier than in East Anglia or the South Eastern Counties, and this is perhaps the reason why the Somerset windmill never grew into the magnificent and elaborate machine found elsewhere.

There are several suggestions why the decline in the South West was earlier than elsewhere, but there seems to be no single solution. The usual story of English windmilling decline is the three mortal blows round about the 1870-1880s, consisting of bad harvests, the introduction of Free Trade in 1875 and the dumping of the U.S.A. wheat glut. There was a series of wretched harvests between 1874 and 1882, particularly in 1875, 1876 and 1877, and when Free Trade was introduced, wheat, fresh from the virgin prairies, was hustled over by the recently introduced steam locomotive on land and steamships on the sea. This encouraged the setting up of steam flour mills close to the wheat ports, and the improving transport to distribute the flour administered the coup de grace.

In Somerset the pattern was somewhat different, a definite and progressive decline in windmills having set in about 1840, perhaps having something to do with an emphasis on pasture rather than on cereal growing. Railways were soon making the transport of milk to London a big factor and other things were forcing the clearer division between pasture and arable. Thus the decline in wheat acreage set in much earlier in Somerset, where the peak year for wheat was 1868, as against 1874 in Norfolk, Suffolk and Essex. Also this diminution of wheat acreage set in faster and on a bigger scale in Somerset, where the loss of acreage between 1866 and 1901 was as much as 56%, as compared with only 32%, 30% and 39% respectively in the above mentioned counties.

The advent of mechanisation, the steam traction engine and the threshing machine 'delivering pulse straw and chaff separately', gave enormous advantages to the Eastern counties where the absence of natural obstacles, such as steep hillsides, waterways, deep valleys and so on, compared so favourably with the little fields of Somerset where, for instance, steam ploughs could never have operated economically. But this factor cannot have become really dangerous so early, nor, in fact, much before the fatal roller mill arrived, and in any case, labour being so much cheaper in the south west, mechanisation would not have been so relatively advantageous there.

Also there was no attempt made in Somerset to take advantage of the advances in windmilling technology of the late 18th to early 19th centuries, such as the fantail for automatically winding the mill, and the shuttered, self-regulating sail. In other counties roller plant was successfully installed in watermills in the 1870s, giving them a new lease of life, but this does not seem to have happened in Somerset. Certainly watermills like Cheddar and Banwell put several Somerset windmills out of business, those on the Isle of Wedmore, for example, and this led to the assumption locally that improved watermills had a lot to do with the demise of the windmill. Certainly the predictability of water supply made them much more viable in an age when hours of working were based on the clock rather than on the wind, but the success of the mills at Cheddar and Banwell was probably as much due to the towns each gaining railway stations about this time and both being served by a radius of roads as to the improved waterwheels and milling machinery.

The South West countryside underwent a terrible depression between 1866 and 1900; cornland was laid to grass, cottage industries disappeared and migration, both to the towns and abroad, depleted the land of labour. Although the countryman generally suffered this black period, it was worse in the south western counties. Although this earlier decline must have softened the blows of the 1870s, for less would have been dependent on corn-growing for their livelihood, it cannot be urged that this curtailing was farsightedness by Somerset. On the whole the answer is much more likely to have been a social one, possibly the increasing importance of such foodstuffs as bacon, meat, cheese and other dairy produce, to the production of which Somerset was better suited. Also it is possible that the repeal of the Corn Laws in 1846 had some selective action unduly penalising the county, perhaps the initially small U.S.A. exports came into south west ports. Whatever the reason for the early decline, it advanced very rapidly indeed after the introduction of steam power and the roller mill. A mere seven Somerset windmills lingered into the 1900s, four of which relied at least partly on auxiliary engines for power.

Only two remained when the Kaiser war commenced, which understandably did something to revive windmilling in some areas, and the last windmill to work in the county was Ashton Mill, finally ceasing work in 1927.

Discussing the last days of the Somerset windmill, a miller once said: 'To my mind the rot set in for wind and watermills in general, first with the coming of the steam engine which made the roller system possible, then the railway brought their end nearer. Then the telephone and the motor lorry finished off both wind and watermills. When the group of windmills round Allerton were all working, each miller went to Bristol or Bridgwater markets to meet the corn merchants to buy his needs and arrange for it to be put on rail for Axbridge or Cheddar. Then we had to haul it by horse and waggon to the mill, grind it and deliver the meal to the farmers, a job which sometimes took several days or a week. Now the farmer just takes the telephone, telephones the port mill, gives his need and the next he knows the lorry is in the yard with the goods, if not the same day as ordered the next day for sure. Of course the war [the 1914-18 war] brought quite a lot of watermills and some windmills back into production for the time being, but the end was only put off temporarily. The port and big roller mills made whiter flour and the stone mill could not compete'.

This is undoubtedly an accurate summing-up but can be amplified a little. One of the heaviest expenses of the miller, whether of a country mill or a port mill, was transport. A two-horse wagon would have a maximum possible daily mileage of about twenty-five miles, carrying at the most perhaps a ton and a half. The port mills, therefore, primarily threatened only the areas within ten to twelve miles radius by road, or those well served by rail. A windmiller outside these areas need not yet as fear competition unduly because, of course, his own delivery area was well within the horse-drawn capabilities. Powered transport by road even in fairly early days could carry two to three times the load of a two-horse wagon and could certainly perform three to four times the possible daily mileage. This meant the port mills could now deliver white flour even to the miller's nearest customers cheaper, because of lower transport charges, and more quickly than the miller could guarantee to do. Even then he could have perhaps claimed a certain share had a sizeable number of customers preferred his produce. This was not so, for many millers took insufficient trouble to produce a thoroughly cleaned product and, although his price for flour as produced might be less, it was generally a less attractive proposition compared with the product of the roller mills, in an age when a physically measurable purity was desirable and could be more easily obtained in the carefully controlled environment of the modern mill.

3 DISTRIBUTION AND DESIGN

The situation of the windmilling areas in Somerset is very interesting and a definite pattern of siting evolves if the known mill sites, as listed later, are charted on a relief map of the county. The majority of sites are found to be in the relatively small area south of the Mendip Hills and north and east of the River Parrett and the area known as the Somerset Levels, about one fifth of the county, claims nearly two thirds of its windmills. A number of sites are found on the edges of the higher ground surrounding the levels and moors, but to the north east and far west of the county sites are rare.

From present evidence the sites fall roughly into three groups:

1. Low island-like hills, where there is slope but insufficient water for power, such as Wedmore, Middlezoy, High Ham and Baltonsborough.
2. Flat moorland, where although water abounds there is insufficient slope to make it usable, such as Kenn, Worle Moor, Watchfield, Edithmead and Chedzoy.
3. Lofty outposts, where for differing reasons, the same conditions obtain, such as Uphill, Hutton, Moorlinch and Worle.

The explanation is, then, that windmill sites abound in districts appropriate for milling but unsuitable for watermills, which latter clearly formed the first choice, for West Somerset, Frome, Taunton, the valley of the Isle and the slopes of the Mendips all abounded with watermills. The best region for corn-growing was stated to be the east of the county, away from the "watery vapours' of the Channel, which affected ripening,[1] and this area was well-served by streams and, consequently, watermills, but on ridges of higher ground rising above the low-lying moors, which often flooded in winter months, windmills were well suited to supply the milling needs of the agricultural communities which grew up on these small tracts of land. In essence, the Somerset windmill was always a local, country, mill.

So far as is known, the Somerset windmills were all either post mills or tower mills, depending on the date of building or re-building, but possibly to some extent on the local availability of timber or stone. Some parts of the county are famous for their building stone, such as Ham Hill, Bath, Dundry and Doulting, whilst in other parts, Bridgwater for example, stone was harder to come by. Henry Corder, writing in the *Somerset Countryman* even went so far as to declare that the county mills were 'mostly built on or near formations of blue lias—of which stone many are constructed'.[2] The actual nature of the stone used to build the towers which have survived for comparison

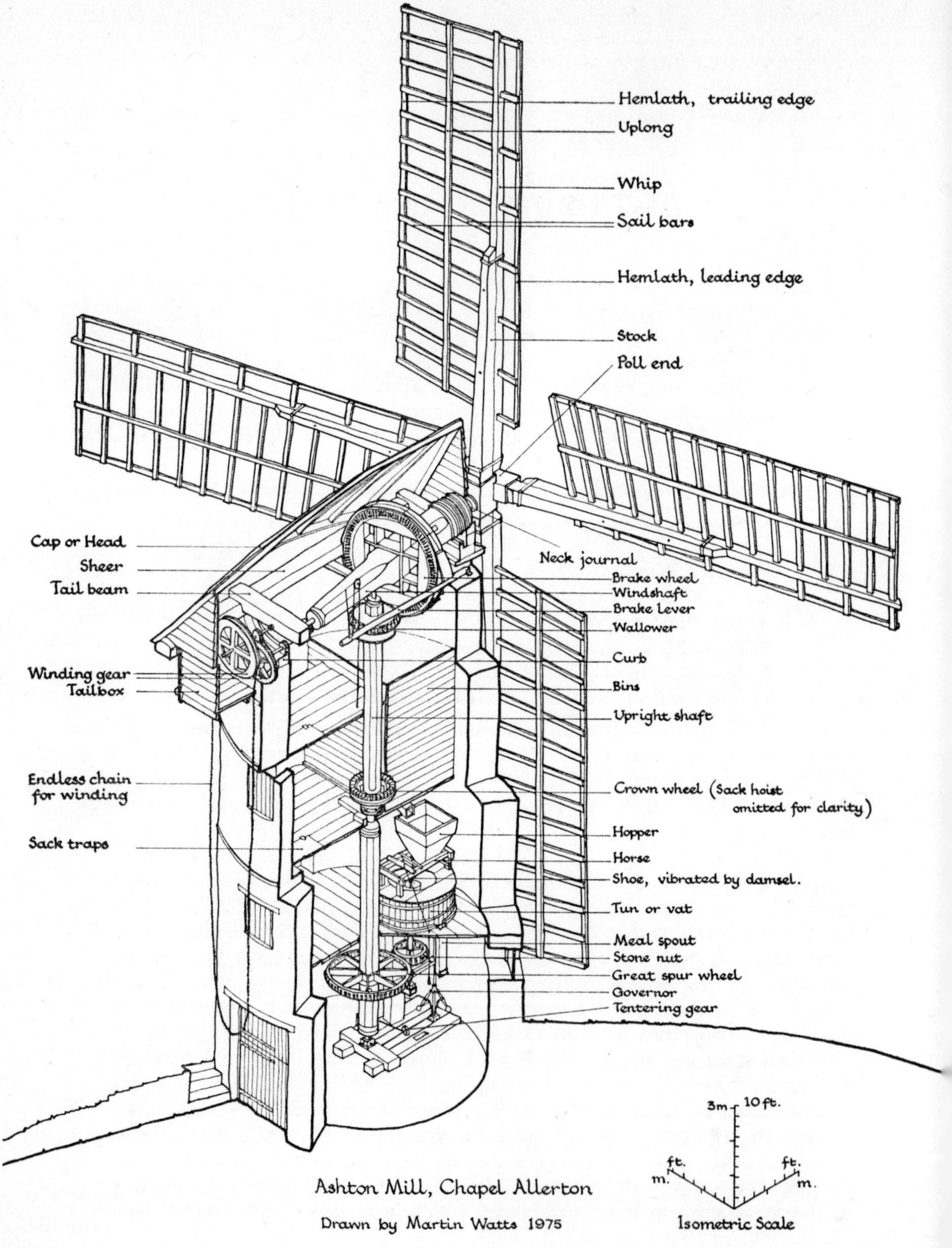

Ashton Mill, Chapel Allerton

Drawn by Martin Watts 1975

varies more than Corder suggests, but it is possible that the tendency for the lias stone to be worked easily into square blocks has in some way influenced the form of the towers built.

Of timber-built post mills in the county little is known and it is doubtful that lack of stone had any influence on the siting of these mills, although the defined post mill sites do appear often to be a little remote from the stone areas. Obviously derelict stone towers have outlasted derelict timber post mills and it is therefore not easy to draw any definite conclusion from the remains and sites identifiable. In short, post mills did work comparatively late in the stone country and tower mills were built on the moors. Also there is evidence that tower mills were built on former post mill sites, as at Ashton Mill, or close to a former post mill, as at Shapwick and Watchfield. The last post mill to have worked in the county was probably Blew's Mill at Edithmead and it was possible to see the ring of sunken stones surrounding the mill mound which probably formed the foot grips to give the miller purchase when he leant against the tailpole to turn the mill into the wind. Although perhaps this was common procedure for Somerset post mills, it is the only site where such evidence was found. The post mill consists of a box-like timber framed and clad body, carrying the sails and the machinery, which revolves around a vertical post so that the sails may be turned to face the wind. The post itself is supported by four diagonal quarterbars which are fixed at the bottom to two horizontal timbers at right angles to each other, the crosstrees. As has been stated, these timbers were sometimes buried in a mound in Somerset. The drawing on an estate map of 1768 of Week, Curry Rivel,[3] is possibly a cartographer's impression of a post mill, either with its substructure enclosed in some form of round-house, or else surrounded in masonry to give stability to the body above.

The Somerset windmill of which most is known in terms of form and mechanical layout is the tower mill. The tower mills of Somerset were generally smaller, both in stature and output, than tower mills elsewhere in England, being the result of different influences and traditions. It is possible that in areas of good building stone, both in this country and Western Europe, the tower mill developed quite early on as an alternative to the post mill, but in some cases apparently never developed beyond the mediaeval tradition of working only one or two pairs of millstones. Somerset tower mills, from present evidence, appear to have been built generally between about 1730 and 1830, and where records exist or remains survive appear fairly typical in form. The towers themselves are generally about 25 feet high (7.6m), with an internal diameter of 12 to 15 feet (3.7m-4.6m) and the masonry wall, some 2 to 3 feet thick (0.6m-0.9m) either parallel-sided or with a straight or slightly convex batter towards the top.[4] Some of the towers were built into mounds, as at Ashton Mill and Shapwick tower mill, some with a circular artificial earth and stone base around, from a few inches in height, as formerly at Watchfield, to several feet as at Stembridge Mill and Weare. This gave a working platform for the miller to adjust the sails from and prevented animals straying too near to the sails when the mills were working. The tower mills at Curry Rivel and Pibsbury are known to have had timber stages around the tower, but both were of unusual height for the county. Another typical regional feature was a short flight of stone steps built up to the opposite doorways at first floor level, and the doorways were placed opposite each other so that one would always be clear of the turning sails.

All Somerset windmills had four sails and all, with the exception of a pair of shuttered sails on Ashton Mill in its last working years, were common or plain sails, set with canvas. A feature on several mills was the broadness of the sails and in some cases the cloth appears to have been rolled around longitudinal bars which were held out by pegs at each end of the sail frame or rolled back as required to shorten sail. In latter years it was more common to have wooden boards placed longitudinally along the leading edge of the sails, but canvas was set on both sides of the whip, the backbone of the sail frame, on some mills. The general direction of rotation of the sails, as seen from the front of the mills, was anti-clockwise.

The cap and sails of the tower mills were turned to the wind by various methods. The cap, typically of gable shape and thatched with timber boarded front and rear gables, turned on a track or curb fixed to the masonry at the top of the tower and gearing contained in a tail-box under the rear of the cap drove to a gear ring on the tower curb. In several mills the cap was originally turned from within by a winch or windlass, and later by an endless chain hanging to the ground from the tail-box, which saved the clamber up into the mill and subsequent procedure. At Vale Mill the last miller is said to have sometimes sat on the sheers in the back of the cap and pushed it round with his feet on the curb. The same miller used to travel to Bristol Docks to obtain old sails from ships berthed there for use on his windmill. Kenn Mill is said to have been winded by tailpole, a braced lever extending from the cap to the ground where the miller could turn the sails into the wind in a similar manner to winding a post mill.

It has been stated that it was customary for millers to brake their mills with the sails set in the form of a St Andrew's Cross on Sundays and holidays. The answer is that in Somerset some did and some did not; one miller recalled that he set them 'prettily' if he thought about it. In a camp near Horsey Mere in Norfolk, very flat countryside from which could be seen some seventeen windmills, including drainage mills, it was related that Horsey Mill itself was used for smuggling and, by arrangement, if any one of the outlying mills altered its sails from a St Andrew's Cross to a St George's Cross, it meant that the Revenue were sighted from the mill in question. Also a recollection that a Signals' Sergeant in the Great War noticed that the sails of a derelict but strangely intact windmill in a French village, although normally fixed in the St Andrew's Cross position were on this occasion moved to the St George's Cross position, and, as a result of his observation, a firing party was evidently detailed in due course, implies that there is possibly some tradition among millers which is, or was, international.

The sails in Somerset tower mills were mounted on a timber shaft, the windshaft, which was inclined at an angle towards the rear of the cap, and was supported on a neck and tail bearing. These bearings were often made of hard limestone, and iron straps and gudgeons were let into the wooden shaft to form a bearing surface. The stocks of the sails were either morticed through the timber end of the shaft, as at Heath House Mill, or wedged into an iron poll end, as at Ashton Mill after 1900. Sometimes the stocks were braced to each other close to the poll end, as originally at both Ashton and Stembridge Mills and Walton Mill. At Stembridge Mill the restored sails are not of the usual construction of whips carrying the frames being bolted onto the stocks, the sail frames being built on to timber backs which are fixed to the windshaft in a channel-shaped cast iron cross.

The primary drive wheel on the windshaft is the brakewheel, on the rim of which the contracting timber block brake acts, and the surviving brakewheels in Ashton and Stembridge Mills are both of clasp arm construction, timber arms supporting the rims being boxed around the shaft. That at Stembridge Mill, however, came from an earlier mill and shows that it was once supported by compass arms, timber arms which were morticed through the windshaft, an earlier practice in gear wheel construction. The remains of the windshaft at Woolavington also show that it once carried a compass arm brakewheel.

In some instances the drive to the millstones was taken directly from the brakewheel, the stones being over-driven. This is known to have been the case at Ashton Mill, before 1900, Heath House, Watchfield and Woolavington Mills, and probably other windmills in the county where only a single pair of stones are recorded. The drive was taken from the brakewheel by a gear at the head of an iron shaft, the quant, to the top or runner stone. In Ashton Mill, again before she was refitted extensively in 1900, the stone nut driven by the brakewheel was a lantern gear, an early form of gear consisting of staves held between two wooden discs. Where more than one pair of stones were driven this was achieved by spur-gearing from an upright shaft in the centre of the mill tower. At the head of the upright shaft the wallower, driven by the brakewheel, was mounted, and the shaft carried a spurwheel from which the drive to the stones was taken by stone nuts. If the spur wheel was mounted above the stones they were overdriven, as at Stembridge Mill, and if below, underdriven, as at Vale Mill. Ashton Mill represents in her present form this latter drive, although to only one pair of stones, the result of being re-fitted with gear from the two-pair tower mill at Moorlinch. It is said that underdriven stones are much easier to dismantle when they have to be re-dressed.

French Burr millstones were commonly used for wheat milling, and several have survived in an ornamental capacity when the rest of the milling gear has gone. Heath House Mill worked with a single pair of 4′ 6″ diameter (1.37m) French stones, which are all that survive on site. Other stones used appear to be of local origin, dolomitic conglomerate from the Mendips and possibly sandstone from the Quantocks. At Broadfield Mill, Felton, and Ashton Mill

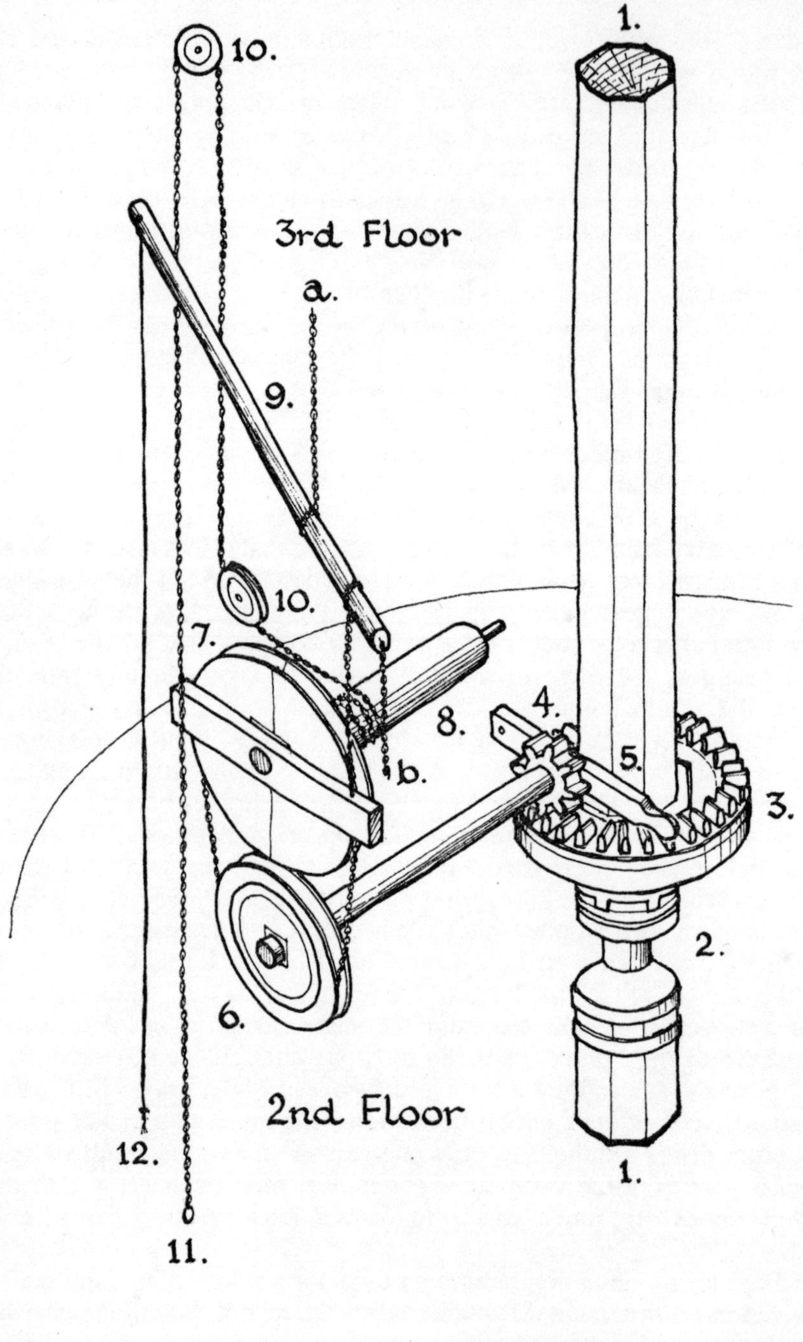

Ashton Mill, Chapel Allerton: Sack Hoist Drive

local stones, both probably from the Mendips, survive, built into the ground. The Ashton Mill stone is 4′ 3″ diameter (1.30m) and from this size to 4′ 6″ (1.37m) seems to have been the common size for millstones in Somerset windmills. Ashton Mill last worked with a pair of 4′ 0″ (1.22m) diameter cement composition stones, a 20th century innovation.

The gearing in Somerset windmills was predominantly of timber, there being little evidence of the sophisticated, precise iron mill-work found in many of the county's watermills. Again it is only in the 20th century re-fit of Ashton Mill that iron machinery has been introduced, a solid cast iron wallower superseding a timber one at the head of the upright shaft, and both the spur wheel and stone nut are iron wheels with morticed timber cogs. Even the wooden brakewheel has had its timber cogs cut off flush to the face and iron teeth bolted on in segments, probably for ease of maintenance. The mill-work in Somerset windmills was almost certainly the work of local carpenters and blacksmiths; one John Latcham, a village carpenter, worked on Heath House Mill, and Smallwood of Worle was the local millwright for Vale Mill late in the 19th century. Millstones were either dressed by the millers themselves or a contracted stone dresser, and Cooper of Cheddar is known to have dressed stones at Heath House and Vale Mills.

ASHTON MILL (*see facing page*)

1. Upright shaft
2. Coupling connecting upper and lower shafts
3. Sack hoist crown wheel (32 cogs)
4. Sack hoist pinion (12 cogs)
5. Lever to engage/disengage pinion
6. Lower chain drive wheel
7. Upper chain drive wheel
8. Chain winding drum
9. Lever, fixed at a. timber below curb
   b. edge of grain bin
10. Chain guide sheaves
11. Sack chain
12. Control rope

The sack hoist is put into gear by lowering the pinion, 4, to mesh with the crown wheel, 3. This turns the lower chain drive wheel, 6. To engage the hoist the rope, 12, is pulled which lifts the upper chain drive wheel, 7, and tautens the chain, so driving the winding drum. The sack is attached to the end of the chain, 11, and passes through the mill via the sack traps in each floor, when it has reached the required floor the rope is released and the chain wheel, 7, drops, disengaging the drive. The whole can be put out of gear by raising the pinion and wedging up the lever, 5.

In both Ashton Mill and Vale Mill sack hoists were driven from a crown wheel on the upright shaft, that at Ashton being a very neat arrangement engaged by a slack chain on two wooden wheels. At Stembridge Mill there is evidence of a hoist, but the gear is incomplete. Dressing machines, to clean and grade the meal after grinding, existed in several mills, but only a wire machine has survived in Stembridge Mill. Ashton Mill at one time had both a bolter and a wire machine, and Worle Mill was advertised in 1870 as having 'two dressing and smut machines'.[5] The storage of grain in Somerset mills was apparently restricted to sacks, for none appear to have been built with a bin floor. Ashton Mill had three large bins built in after 1900, but the general arrangement appears to have been that corn was emptied into the hopper over the stones as required, and this does indicate that the mills were supplying a local demand, each farmer bringing a specific amount of grain to be milled, rather than being flour factories.

The majority of windmills known in the county were corn mills, but some references do exist to water pumping mills. The Somerset Levels were drained methodically for many years, but windmills used in this capacity were few. Common Moor, near Glastonbury, was drained by 1722 and there is a reference to an 'engine to throw the water into all the Ditches of Common-Moor', which could well have been wind-powered. At Bleadon a windmill built about 1613 was used to keep the sea from over-flowing reclaimed land, but these references are the only ones found which illustrate this use of wind-power in the county.[6] When the larger drainage schemes were initiated in the 18th century, the majority of drainage appears to have taken place using gravity to make the water flow away. At Pawlett Hams a wind-pump was used to pump fresh water across the River Parrett by pipe in the late 19th century and this is described in the list of sites.

A final use of wind-power in Somerset in a light-industrial capacity was at Wellington, where a tower mill apparently once stood near the middle of the town and was used for grinding oak bark for the tanning industry, but this seems to have been an isolated example.

Above: Post mill carving on the metope frieze of the Circus, Bath, 1754-8.

Left: Bench end in Bishop's Lydeard parish church, early 16th century.

Above: Bench end in North Cadbury church, 1538.

Left: Restored 15th century wall painting of St Christopher in Ditcheat parish church.

Left: The overgrown ruins of Brockley windmill tower, looking south.

Below: Quarterbar and part of cross tree uncovered in March 1971 in a mill mound near Bridgwater.

WATCHFIELD MILL FROM A POSTCARD OF ABOUT 1905

Note the low platform around the tower base and the weighted endless chain on the winding gear hanging from the tail-box of the thatched cap.

WATCHFIELD MILL IN DECAY IN THE LATE 1920s

The sail frames have been removed.

4 THE MEN WHO WORKED THEM

The first windmill sails I ever saw turning were those of Ashton Mill. My parents used to take a house for holidays that was situated up on the Mendip Hills not far from Cheddar. From there, looking across the flat valley of the River Axe and glimpsing through a belt of trees, one saw a queer, jerky movement. I say queer because Ashton Mill, like most Somerset windmills, was a compact structure and, being partly hidden amongst the trees, was only detectable by the movement. This was some years before the Great War and it was not until I was at school not so very far away that I contrived to find and visit the mill.

Structurally it was not very imposing. To me Lancashire tower mills were massive and fortress-like, Norfolk and Kent mills were structures of great beauty; those around the Wash were almost exotic with their incredibly lofty towers and abnormal numbers of sails and post mills, often painted white, seemed toy-like, but the Somerset windmills were not often beautiful apart, perhaps, from their thatched caps and often their setting, Somerset being a well-wooded, well-watered pasture county, but nevertheless they conveyed a much greater impression of mill and miller being component parts, rather than the ' captain on the bridge ' impression others often gave.

It is of some of these millers I wish to write. If millers in other parts might on occasion be sophisticated almost ' city gent ' types, reading, no doubt, the *Financial Times* in between whiles, they were never like this in Somerset. All those I knew were simple men wedded to their craft, opening up only to a kindred spirit and having a personality shared only perhaps by thatchers and blacksmiths. It is not just being ' rural ', as it differs a good deal from that exhibited by countrymen dealing with live animals, shepherds for example, perhaps because their cares are perpetually changing whereas the miller becomes identified with his mill, the smith with his smithy and the thatcher with his peculiar tools and materials.

The Somerset millers were to me, then, the ' combat troops ' of milling. Nor is this an inappropriate description. There were not a few who had lost one or more fingers and many who had lost a hand or an arm, whilst there were few running repairs or even major breakdowns that called for commercial attention. 'Apple and hornbeam cogs ' used to be advertised on a firm's sign alongside the Great Western Railway near Bristol's Temple Meads Station and, with the aid of this firm, the village blacksmith and carpenter, plus the ingenious grafting of material from abandoned wind- or watermills, their mills were rarely out of action very long, provided the ever-present risk of fire had not materialised.

When I knew it first, Ashton Mill belonged to Mr John Stevens and it was worked for him by a sequence of millers, the only one of whom I knew really well being Tom Petheram, although he had long left this mill when I got to know him. Mr Stevens was a keen and enterprising miller. He was indirectly connected to another famous Somerset milling family, his sister having married Joseph Loader of Moorlinch Mill, and his daughter later married James Loader of the same family. John Stevens himself was born within 500 yards of Moorlinch Mill and it was this mill that aroused his early interest and where he gained his milling experience. As related later, Mr Stevens visited the Clayton windmills in Sussex to see for himself how shuttered sails operated.

Tom Petheram was fascinated by Ashton Mill as far back as he could remember. His grandfather was a thatcher and Ashton Mill, like other Somerset windmills, had a thatched cap. Thatch must have been widely used for roofing mills at one time because one encounters old agreements such as ' tenants were bound to the lord's mill to pay the fortieth corn and to maintain the wall and thatch of the mill ' and I have seen an indenture of 1676 requiring the tenants to ' maintain the mill with wall and thatch '. Tom used to go along with his grandfather when thatching and, having once seen the mill, could not be kept away from it. His grandfather's remuneration for thatching the mill included so many bacon dumplings, made by boiling bacon sandwiched between dough in a cloth. When the time came for Tom to start work, at ten years of age, that is therefore where he went. This would have been about 1888. He was running the mill by the time he was eleven years of age when his wages were 2/6d per week, plus meals, and his hours ' from windrise to wind-set'. As the charge for grinding a sack of wheat was 1/6d and out of a sack of 224 lbs (101.6 kgs) 50-56 lbs (22.7-25.4 kgs) of bran would be obtained, the rest coarse flour, and the mill would probably grind up to four sacks in an hour, he fully earned his keep. Early on, Tom was nearly killed by being struck on the head by one of the sails when hurrying to the mill which, briefly unattended, had begun to race because of a sudden storm.

Windmills ' running away ' were not, of course, uncommon, but these Somerset mills, although not lofty, were very powerful and Ashton Mill, although only about 150 feet (46m) above sea level, had no ground of any elevation between it and the Atlantic. Furthermore, the mill drove only a single pair of stones and a simple brushing dresser and, if running out of corn in full sail with a sudden gale, the danger of fire must have been very real. Tom never forgot once when seeing such a storm coming up he was about to stop the mill to take in sail. Mr Stevens, seeing this, said ' Don't stop her; I'll stay until the storm blows over '. Tom told me ' We had only just got inside when the storm struck us. Did it blow ! and did that old mill run ! I put the brake on and sat on the lever, but it only seemed to make her go faster. We were almost choked with smoke caused by the friction between the wooden brake and the wooden wheel. Sparks were falling like rain and the mill rocked from top to bottom. Believe me, we had to hold fast to anything we could get hold of to keep our feet. I can remember that day quite well (this was sixty years after) and the look on Mr Stevens' face as he said " My God, what shall we do? " But the storm passed, the wind turned and we managed to stop the mill. I never saw the sail rolled in before or after as Mr Stevens rolled it in that night '.

Evidently it impressed itself on Mr Stevens too, as he journeyed up to Sussex to view the famous windmills Jack and Jill, one of which was fitted with automatically adjusting sails with spring-controlled louvres which opened if the wind proved excessive.

Tom Petheram left Ashton Mill in about 1897 to work in Willett's Water Mill at Banwell, once a famous mill in a most picturesque setting, the mill-pond always with swans upon it, forming the centre of the village with great overshot wheels flanking the pond. The whole setting was clumsily destroyed when the water supply was purloined by a distant water board. About this time, Spillers and Baker, who pioneered steam corn mill hereabouts, attracted Tom to a new mill at Cardiff and thence to Gateshead. It is hard to imagine a greater contrast than that between Ashton and Gateshead. The densities of population must have approached the minimum at the former and the maximum at the latter, whilst the four floors and single pair of stones at Ashton contrasted similarly with the thirteen storeys and 80 sacks of flour per hour at Gateshead.

Tom Petheram died on 5 November 1953. He knew that at that time all efforts to preserve Ashton Mill had failed and when he last visited it he was

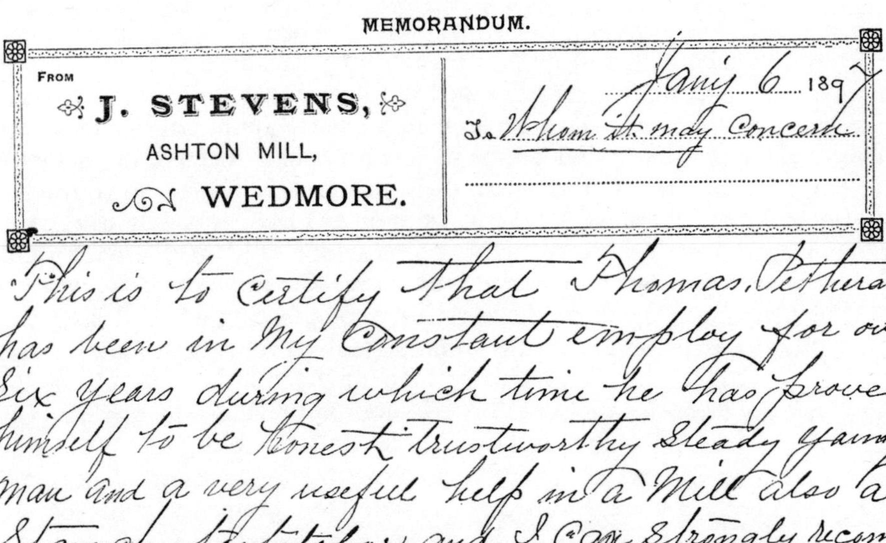

MEMORANDUM.

FROM
J. STEVENS,
ASHTON MILL,
WEDMORE.

Jany 6 1897
To Whom it may Concern

This is to certify that Thomas Petheram has been in My Constant employ for over Six years during which time he has proved himself to be Honest trustworthy Steady young man and a very useful help in a Mill also a Staunch teetotolor and I can strongly recommend him to anyone
John Stevens

Miller's testimonial

saddened to see its ruinous condition. He would be happy to learn that public interest was at last aroused and his mill preserved.

The head miller at Banwell watermill was a Mr Tucker, whose brother William worked Heath House or Westfield Mill, Wedmore, from about 1876 until he died in 1902, a few years after the mill stopped working by wind, and a Mr W. F. Duckett took over, whom I got to know in 1922. Mr Tucker's daughter had kept a school of about twenty-five pupils at the mill house and one of these pupils was a William Duckett. He too it seems was curiously drawn to the windmill at an early age and claimed that Westfield Mill was the best of many in this part of Somerset, although when I told Tom Petheram this he said, ' Ashton Mill was often grinding steadily in winds that would not stir the sails of our local rivals—Heath House included '. William's father was Michael Duckett, the owner of the first mowing machine in Wedmore about 1873.

Westfield was for much of its time worked by members of two famous Somerset milling families, Wilkins and Spearing. George Wilkins died in 1810 and on 7 June 1833 John Wilkins was struck dead by lightning in the mill. Whilst in the Tucker's ownership, John Spearing from Watchfield Mill was chief miller for many years. Here he lost three fingers from one hand when pulling the cog into gear; the fingers were put into a bottle and buried with his wife only six weeks before his own death. John Spearing had three sons and four grandsons, all millers in this county. One of his grandsons, well known to me, was William Spearing, born at High Ham Mill, which he could work when he was ten years old. When he was thirteen, he caught his hand in the spur wheel at Paradise Watermill, Low Ham, crushing it and the lower part of his arm which was amputated at the elbow. When I knew him he was at Isle Brewers Water Mill and despite his infirmity was able to dress the millstones himself. The mill extracted a further tribute when his son was drowned in the pond of this mill. The Spearings and the Wilkins intermarried and between them must have provided the millers for over a dozen of the Somerset mills, but, if milling was in the blood, there was no other career to consider.

Worle Mill, on Worle Hill near Weston-super-Mare, was last worked by William Rogers who combined, as was almost universal in this county, the occupations of miller, corn merchant and baker. The mill ceased work about 1870, but Mr Rogers' nephew, Samuel Curry, was fascinated by it as a child and recalled his experiences vividly and excitedly seventy years later. Once the mill was idle for an exceptionally long time for lack of wind and Mr Rogers and his three sons mounted twenty-four-hour vigils for a wind. One night when it was Mr Rogers' own watch, the wind did spring up. ' It was my uncle's turn to go up and let the brake go. Now there was an accident just as he let go the rope, it caught in his leg or foot. He fell and broke his leg. He was a very brave man; now, with his broken leg, he managed to get to the doctor's at 2 a.m. It was a very rugged road and steep hill to get down. (This is an understatement; I knew the track well). Well, he got down there and had the leg set and home again before his people knew anything about it. Not bad was it? ' Samuel Curry, who told me this, was himself connected to yet another Somerset milling family, the Emerys. Between 1700 and about 1840 twelve individuals of this surname were milling in ten different Somerset wind- or watermills.

Yet another was the Hookway family who must have been millers for at least three generations, but I never met any of them to my knowledge. One Robert Hookway was miller at Stone Allerton, Weare Mill, in the early 1860s, at about the same time as John Hookway was miller at Pibsbury Mill. This latter mill was built by William Wheller and members of the Wheller family were millers in Huish Episcopi in the early years of this century.

The millers, whether owners, tenants or employees, were generally very loyal to their mills, each claiming some special virtue for their own. Whereas around Stoughton five windmills were in sight of each other's sails, keen observation was kept in a period of calm to see which would respond most readily to the long-awaited light wind and I have been privileged to hear counter-claims as to which windmill would go into action most readily. Masters too varied equally in their determination to utilise these winds, some millers working wind hours rather than daylight hours and one suspects that the not infrequent change of master by moving to another mill had something to do with this. Strict Sabbatarians were popular with one employee to my knowledge, who usually supplemented his earnings at the mill by doing a bit of rabbiting on this day of the week.

Another miller who had a pretty daughter seemingly had no difficulty in recruiting milling assistants, who again were replaced equally frequently. One of these coming out of the mill with his eyes on Betty was struck unconscious by a blow on his head from the rotating sail, and only just escaped with his life. He told me this many, many years later, but said he had thought the girl was bringing him a message. In the case of automatically luffed windmills where a change in wind direction might have unexpectedly brought the sails to the door side of the mill, this must have been a great danger. This would not apply to Somerset windmills, as none were equipped with fantails, but I have spoken to millers who have had this experience, usually, I suppose, in converted mills previously winded by hand, as this would be unlikely to be a possibility on most windmills with fantails.

5 SOMERSET WINDMILLS AND SITES

This list of known windmills and sites cannot claim to be exhaustive for, as has been said, more detailed work has still to be done in many Somerset parishes. The purpose is to record such information as has been already gathered from numerous sources and references, arranged alphabetically under the present parishes. All site grid references are given to six figures where possible and the prefix letters of the National Grid which cover all the Somerset windmill sites are ST.

ABBAS AND TEMPLECOMBE *approx. 705 218*

A field called 'Windmill Shards' is recorded in the Tithe Award.

ALLER

A close called Windmill Ground in Aller Moor was mentioned in 1796.[1]

ASHCOTT *1. approx. 442 377*
2. approx. 437 375

A windmill is marked at Ashcott on Strachey's map of 1736 and Bowen's of 1760. 'Millslade' is recorded at the first reference on White's survey of Walton of 1784[2] and Millslade Farm exists today, with a mound to the east. 'Mill Green' is recorded in the Tithe Award at the second reference.

ASHILL *approx. 315 168*

Windmill Hill is marked on the OS maps from the early 19th century and 'Windmill Lane' and 'Windmill Hill Farm' appear in the Tithe Award.

BALTONSBOROUGH *approx. 549 351*

Windmill Hill is recorded on the OS maps and in the Tithe Award.

BAWDRIP *approx. 347 397*

A windmill is marked on 17th century maps near the present King's Farm, notably on Newcourt's map of 1660 and Ogilby's road map of 1675. The site has apparently been obscured by a railway.

BEDMINSTER *1. approx. 592 714*
2. approx. 583 701

A Windmill Hill still exists in this former Somerset parish, now a suburb of Bristol, and a windmill is marked here on Benjamin Donn's 'Map of the Country 11 Miles around the City of Bristol' of 1769. Donn also shows a windmill at the second reference, apparently in connection with a watermill.

BERROW *294 523*

A windmill is recorded in Berrow in the early 1330s,[3] and on an estate map of 1773 a tower mill is shown amid the sandhills on the seaward side of the path connecting the church and village.[4] It is also marked on Day & Masters' map of 1782 and is said to have ceased work about 1815, after which it was used as a dwelling house. At some time during this latter period it was said one occupant intended taking the life of the other and got as far as digging a grave for the victim. The plan was, however, discovered in time to avert the crime, but no-one was prepared to live in the mill afterwards and it gradually fell apart. Local recollections date the tower's final demolition as sometime 'in the nineties' and in about 1930 the remains consisted of a ring-shaped mound, some grass-grown banks and a gateway.

BICKENHALL *approx. 282 180*

The Tithe Award records 'Windmill' at the above reference and a Windmill Cottage is still in existence.

BISHOP'S CAUNDLE *approx. 720 119*

A Windmill Hill exists to the east of Holwell, although these former Somerset parishes are now within Dorset.

BLEADON

A pumping windmill referred to in the early 17th century on one of the reclaimed warths in the Bleadon Level, the estuary of the River Axe.[5]

BRENT KNOLL, South Brent *1. approx. 332 509*
2. approx. 335 516

A windmill is recorded at Southbrent in the early 1330s on the estates of Glastonbury Abbey of value £2,[6] and a windmill is marked on John Ogilby's road maps of 1675. The last windmill to work in the parish was Shackleton's Mill and Ellis Shackleton is recorded as miller in the trade directories for 1861, 1866 and 1875. This mill stood at the first site, on the lower slopes of Brent Knoll between South Brent church and the Dun Cow Inn, and is said to have been a stone tower mill. It is understood to have been pulled down in the early 1900s and the masonry re-used to build a dwelling house nearby. Shackleton seems to have been a bit of a character, from the following information: 'Shackleton was a Quaker who ran into debt and was troubled by his creditors. He would only discuss business matters when employed in his mill. If a creditor called at his private house, Shackleton would tell him this. However, repeated visits to the mill itself met with no response and the creditor reproached Shackleton for breaking his promise. The miller replied "I said I would see thee if thou camest to the mill. I did see thee through the door, and I liked not the look of thee", and stalked off'.

The second site is evidence from the Tithe Award, perhaps the location of an earlier mill.

BRIDGWATER WITHOUT *318 384*

The site of a mound which, when excavated by archaeologist Trevor Miles in March 1971, was found to contain the sub-structure of a post mill. The mound was some forty feet in diameter and about eight feet high, above the level of the field, with a shallow ditch about twenty feet wide surrounding it. The mill timbers were removed and are at present preserved in sand at Blaise Castle House Folk Museum, Bristol. The mound contained pottery fragments of 14th-15th century date. The site is now obscured by an M5 access road.

BROCKLEY *473 661*

'By an inquisition taken at Langport 17 Oct. 20 Hen VIII (1529) it appeared that Richard the son and heir of Humphrey Harvey died 4 Jan. 17 Hen VIII (1526) seized of one third of the manor of Brockley, five messuages, one cottage, one windmill . . . Which premises were certified to be holden of the King as of his barony of Wigmore by knight's service'.[7] In 1630 the miller was one Edward Hill of Brockley who, regretfully, furnished ammunition for those who suggest that millers were not always models of business integrity. He was in fact arrested and charged with the theft of some corn left with him to be ground. He was examined before a Justice of the Peace, Richard Cole, on 12 November 1630 and was apparently so hurt at being thus slandered that he immediately fled the country.[8] A windmill is shown here on Donn's map of 1769 and Day & Masters' map of 1782, but by 1829 the 'ruins of an old wind-mill, erected on the very apex of the down' was all that remained.[9]

The ruins of a tower mill can still be found in Brockley Wood, although the approach is difficult. The tower is in a very decayed condition, of large diameter and with an exceptional degree of batter, although only part of the tower wall stands on the north side. It was locally spoken of as 'the granary' and corn was at one time grown on the adjoining Wrington Warren.

BURNHAM *309 484*

A windmill is marked near the above site by John Ogilby in 1675, possibly a post mill preceding a tower mill, for in 1786 an insurance policy refers to a windmill of brick and timber, valued at £100, in the possession of Thomas Henbury of Burnham.[10] In the 19th century a windmill known as Allen's Mill or Sandway Mill existed on the site of Burnham Brewery, and in 1869 machinery offered for sale by R. W. Shackel of the Burnham Brewery included ' two pairs of famous Mill Stones, Picks, Tools, Cap, and all Fittings appertaining to a Windmill ' with the note that ' all brick and stone to be left upon the premises '.[11] This appears to have marked the end of this windmill, which was worked by one John Allen, brewer and maltster, in the early 1840s. He was apparently preceded by J. Lennard. The site is still part of a brewery.

BURNHAM WITHOUT, Edithmead *337 490*

A windmill is marked in this vicinity on Ogilby's road map of 1675 and also on Bowen's map of 1760. On the first edition OS map it is marked as ' Burnham Mill ' and in the Tithe Award as ' Blew's Mill ', the name under which it became well-known locally. The windmill which stood on a mound to the south east of the main A38 road was possibly the last post mill to work in Somerset, although detailed particulars have been exceedingly difficult to collect, even though a great number of people could recollect their families having corn ground at this mill. No doubt its very easy access contributed to its popularity. In about 1930 the remains consisted of one or two large stones on top of a mound, which still exists, and there was evidence of a ring of sunken stones marking the track round which the tailpole was pushed or pulled to wind the mill. The mill has been described both as an open trestle post mill and as being of wood with a stone base a few feet high, and was last used sometime between 1875 and 1885. After it was pulled down the oak post was converted into a roller for the land !

The last miller, Joe Blew, who seems to have previously worked Watchfield windmill, was something of a local celebrity with his inseparable churchwarden pipe. It is said that ' the rural inhabitants brought their grist in small bags which had been gleaned by hand and left it outside the mill to be ground. Joe Blew trained the male bird of his poultry to peck a hole in the bag to get at the corn and then call to his family of hens thus, " Here ! Here ! ! Here ! ! ! ". This was his own story '. Joseph Blew died on the 23 October 1890, age 70. It is said he was killed on the railway line.

BURNHAM WITHOUT, Highbridge *322 485*

A windmill is marked on Ogilby's road map a little to the south of this site, where a mound is still in existence on the low-lying ground between the B3139 and the railway from Highbridge to Burnham. The mound has been excavated in recent years and found to contain incomplete timbers from a post mill trestle.[12] A cross-shaped depression on the summit of the mound which is still evident is the result of excavation.

BURNHAM WITHOUT, Watchfield *1. approx. 343 472*
2. 348 470

The first site is that of a post mill which is said to have worked until about 1867. In an insurance policy of 1803 a ' Corn Wind Millhouse timber and thatched ' was valued at £30 and ' the standing and going geer, millstones, etc, therein ' at the same amount, and the mill and dwelling and bakehouse nearby were in the ownership of one Ferdinando Hawkins.[13] The site can be verified from the Tithe Award.

Watchfield tower mill, of which the tower still stands, is first marked on the 1817 OS map and worked for about 100 years, until the beginning of the Great War. The tower is well built of coursed lias stone, with a slight convex batter, and the mill had a thatched cap and the usual chain luffing gear and worked with a single pair of over-driven stones. A 4′ 6″ (1.37m) French Burr stone can be seen by the front door of the present house. Although the gear of this mill survived until the 1930s, there is now no other trace on site. Watchfield Mill was worked for much of its life by the Spearing family, although in 1841 the tenant was James Hookway and the owner Sarah Blew,[14] and Joe Blew, of Burnham Mill, Edithmead, was miller until 1861. John Spearing the elder was miller until 1874 when his third son, also John, took over, working the mill until his death in 1893. Frank Spearing, nephew of John the younger, came here from Shapwick Mill in 1894 and worked the mill until it closed down, using wind and steam power.

William Spearing, of Isle Brewers water mill, related that his uncle, John, when miller at Watchfield, once met a racehorse owner from near Bristol then staying at a Weston-super-Mare hotel. In the course of conversation the owner introduced the subject of horse-racing and dwelt upon the ability of a certain horse of his, claiming it to be unbeatable. At this John Spearing mentioned his half-bred mare that he drove in the mill cart and said he would back it against the other for a stake of ten pounds over a mile course of his own choosing. The racehorse owner jumped at this apparent gift of ten pounds and all was soon agreed, time and date fixed and the meeting place to be Watchfield Mill. When the day arrived, John Spearing was working his mill, but went down to the house to meet the visitors and, after telling his man to bring out the mare, took them to the mill paddock where the mile course had been marked out passing round the windmill eight times. The race then started and the miller's mare, being used to the mill, soon reeled off the eight laps. The racehorse, however, refused to pass the still-working mill, and so lost the race. ' That ended in a great spree ' said William Spearing, who had heard the story from his uncle John nearly 50 years before. To the suggestion that it was lucky for his uncle not to have suffered a calm day, the nephew rather thought that, come calm, come wind, his uncle would somehow have come out on top.

The tower of Watchfield Mill, although gutted of its milling gear, has been well-maintained and has recently had a conical thatched roof put on.

BUTCOMBE *517 628*

This is a very fine windmill site, over 500 feet above sea level and offering what has been described as the best view in all Somerset. Although there is no documentary evidence as yet for this site, the mound which can still be seen has been excavated and the positive remains of a sunk post mill found.

The site is in corn country, a good exposed location for a windmill, and a trackway is discernible leading almost to the foot of the mound.

In 1946-7 a trench was dug through the mound and the remains of a windmill which was active in the 13th century and had been built into an earlier tumulus were found. The mediaeval finds, other than pottery, included 140 iron objects, a small piece of lead spilled in the molten state and six fragments of three millstones, of which the largest was originally about 4′ 0″ (1.22m) diameter and 4″ thick (100mm) and was fire-scorched. All three pieces were of sandstone conglomerate with quartzite pebbles, probably quarried on the nearby Mendips. The iron objects included many nails and what is possibly a latch-lifter, and a dozen fragments of sandstone roof tiles were also found.[15]

Mr. Philip Rahtz, under whose direction the excavation was made, wrote: ‘Evidently in the case of Butcombe, a large pit which may have been ‘T’ shaped was dug in the centre of the barrow when the foundation timbers were laid, then filled with the mixed material dug out’. It was his opinion that the windmill belonged to the Cistercian cell of Butcombe Court, and that it finally collapsed.[16]

BUTLEIGH

1. approx. 513 324
2. approx. 506 326
3. approx. 527 325

John Ogilby marks a windmill at Butleigh in 1675, and the first edition OS map shows a windmill at the first reference. This site is not qualified by the Tithe Award, but the two other sites are indicated by field names. The first site is close to the present reservoir. The lack of information even when three sites can be claimed for a single parish gives some indication of the difficulty of defining the exact location of windmill sites and the inconsistency of information which can be gathered.

CATCOTT

approx. 388 383

John Ogilby's map of the road across the Polden Hills of 1675 marks ‘Catgut Windmill’ close to the road, and the site is qualified in the Tithe Award although the windmill was no longer standing. A field called ‘Mill Ham’ is marked in the same Award at 405 401 on Catcott Heath, but it is doubtful if this was a former windmill site as it is in the lee of the Polden ridge.

CHAPEL ALLERTON

414 504

The Manor of Allerton was bestowed by William the Conqueror upon Walter de Dowai and sub-let by him to Ralph de Contiville.[17] The manor passed through Adam de Contiville to Richard, and, on his death, to his son Baldwin de Contiville in 1303. Whilst in the latter's hands, a fine of 1317 specifically mentions a mill, though its type is unspecified. The mill is mentioned in connection with the manor during the 15th century and in 1549 an indenture was made with John Mawdley, a clothier of Wells, for the 're-building of a windmill'. Whether the windmill passed out of the hands of the manor to Mawdley is conjecture, but in 1705 the mill and manor were again under one owner, one John Paine, who was appointed by the Dean and Chapter of Wells to be their gamekeeper in the district some two years later.[18] After Paine's ownership the mill and the manor appear to have been separated and it was perhaps at that time that the windmill came into the hands of the Luckins family, for it was said to have been with them for over one hundred years and by about 1861 it was owned by Paul Wilkins. Ownership passed to his son Edwin in 1866, T. Hooper was owner from 1884 to 1887, and John Stevens, the last windmiller in Somerset, was owner from 1887 to his death in 1938, the mill stopping work in 1927, some 600 years after the first reference to a mill in the manor.

Ashton Mill was built sometime during the third quarter of the 18th century, for it is known to have been standing in 1774[19] but does not appear on Bowen's map of 1760. It is said to have been built on a mound formerly occupied by a post mill and to have incorporated timbers from this earlier mill, one of which was dated 1736.[20] This cannot be proved, but the re-use of timbers was common practice and those at present supporting the stone floor, for example, have definitely been used previously. Ashton Mill was, however, altered quite considerably at the turn of the century and has undergone extensive restoration within the last twenty-five years.

Before being 'modernised' by John Stevens in about 1900, Ashton Mill contained some basic machinery of an earlier type than commonly survives in windmills. The sails were broad with cloth each side of the whip, a form which can be seen in illustrations of mediaeval windmills but had generally been superseded by the 19th century, and the sail stocks were morticed through the timber windshaft. The windshaft carried a trundle wheel which engaged with a lantern pinion mounted on a quant, the lower end of which engaged with an over-driven runner stone. The single pair of millstones were located probably one floor higher than those in the mill at present. About 1900 machinery was brought from the recently demolished Moorlinch tower mill, and Ashton was extensively refitted. The thatched cap, originally turned internally by a windlass, was re-framed, clad in corrugated iron and the endless chain winding gear from Moorlinch Mill was fitted. A cast iron poll end was fitted to the windshaft and two of the broad sails were replaced by Spring sails, the only pair ever to be fitted to a Somerset mill. These provided a degree of self-regulation, while the common sails gave a little more power—a good working combination. It was at this time that the large bins were built in below the cap, the single pair of stones were moved down one floor and the upright shaft and underdrift gear installed. There is no room for more than a single pair of stones in the tower, and therefore the typical indirect drive designed for two or more pairs of stones is connected to a single pair only. The upright shaft, which consists of two shafts connected by a dog clutch, is of pitch pine above and oak below, and carries the cast iron mortice spur wheel from Moorlinch Mill which drives a single iron mortice stone nut. At the head of the upright shaft can be seen the original wooden wallower, with the newer iron one close above it.

A crown wheel provided a secondary take off for the sack hoist, which also was brought from Moorlinch, and the dressing machinery which has now gone. The pre-1900 dressing machines consisted of one bolter about four feet long and two feet in circumference. The fabric cover was in one piece and drawn on like a glove; great care was taken of the cover and it was taken off and cleaned after every six sacks or so of meal. Some bolter cloths were silk, but Tom Petheram did not think that at Ashton was, but from Mr Stevens' care and concern ' it might well have been'. There was also a little brush machine about two feet long, the first quarter of which was clothed with fine wire and the brushes were driven fairly fast. The feed to this machine came direct from the stones through the trough bottom by taking off a hand hold cover. This machine was very satisfactory and preferred by Tom Petheram to the bolter. With this set up and with a favourable wind the output of the mill was from twenty to thirty sacks per day. After 1900 these dressing machines were replaced with a new bolter, with a woollen cloth, and a new or second-hand wire machine, both apparently sited on the second floor. None of the dressing machinery now remains.

Externally, Ashton Mill has a distinctive feature of three iron hoops around the upper part of the tower, which were also fitted about 1900. It has been said[21] that parallel-sided windmill towers are examples of bad design in that the cylinder of masonry is more inclined to distort under the weight of the cap and sails than a battered tower and, if the rack on the curb is not truly horizontal, or truly circular, the cap would tend to jam. Both John Stevens and Tom Petheram, however, maintained that Ashton Mill always ' turned sweetly ', but the iron hoops were fitted during Mr Stevens' ownership, so presumably he was taking no chances. Ashton Mill was originally whitewashed externally, as were several other Somerset tower mills, and had a number of wooden blocks through the thickness of the masonry wall just below the curb. These, and again they could be seen on other local tower mills, were presumably for levelling and fixing the timber curb to, an easier task than to rubble masonry.

After stopping work in 1927 Ashton Mill fell into disuse. That she had been able to work on when all other windmills in the county were unemployed was due in many ways to the foresight of Mr Stevens in up-dating the grain storage and the machinery. Steam power had been introduced at least as early as 1894, being supplied by a portable engine which drove the stones by a belt drive to an external pulley, the wind-driven gear being disengaged, but in the latter years she was reduced to grinding animal feed for local farmers. For over thirty years the mill stood disused, but in 1958 she was restored at the charge of the late C. C. Clark of Stone Allerton by John Scourse of Cheddar, to a specification drawn up by Rex Wailes. Mr Clark was presented with a Windmill Certificate by the Society for the Protection of

Ancient Buildings for his efforts and he passed the mill to Bristol City Museum in 1966. At the time of writing the mill is in need of some repairs, the Douglas Fir stocks which are slightly shorter than those originally on the mill having deteriorated badly, but it is hoped that the mill will soon be put back into a state of good repair.

ASHTON MILL HOUSE

A document of 1712 refers to the 'domus molendinarii' now built in the upper field near the windmill,[22] and Tom Petheram recalled a low, thatched, white-washed house 'very like the mud-built houses at Clewer, about two miles from Ashton towards Cheddar', where he had once lived. The house had three bedrooms, 'two windows in the three rooms. The middle bedroom hadn't a window. A wood block about the size of a brick fitted into the wall to be taken out to let air into the room. There was a good-sized oven at the back of the house where Edwin Wilkins used to bake the bread. He started to take bread to some of the outlying districts with his horse and a covered (Coburg) cart'. This house has now gone, the mill standing in an isolated position to the north of the stone-built Ashton Mill Farm, at the top end of a field called Mill Ground.

CHARLTON HORETHORNE *approx. 674 239*

There is a Windmill Hill here and it is said that the windmill was standing until about 1784,[23] although it is not marked on any of the 18th-century county maps.

CHARLTON MACKRELL/CHARLTON ADAM *516 292*

There were two mills on the estate in 1467 and a windmill on the present Windmill Hill is first mentioned in 1616. In 1626 it is described as 'Mr Thomas Baskett's Windmill', Baskett having died in 1592, and on an estate map of 1757 a tower mill is drawn in elevation on Windmill Hill. This mill had apparently been demolished by 1810.[24]

CHARLTON MACKRELL *1. approx. 530 286* *2. approx. 551 271*

Near Lytes Cary 'Windmill Field' is marked on the Tithe Map, and the second site, at Cary Fitzpaine, is also suggested by a field name in the Award.

CHEDZOY *345 371*

The track leading from Chedzoy to Mount Close Batch traverses a huge ploughed field of three hundred and sixty-five acres, it is said. On the right-hand side of this track, on perhaps the highest part of the field, is a still slightly raised area, despite repeated ploughing, which is called 'Mill Batch' and was the site of Chedzoy windmill. A windmill on the estate of Henry Porter was 'carried away by the volume of wind from the eminence on which it was situated to a distance of 30 feet' on 7 December 1827, and the miller, Joseph Luke, was in the mill when it was carried off but 'received no injury except a few slight bruises'—a fortunate man.[25] This mill is marked on the 1st edition OS map and on Greenwood's map of 1822, and in April 1907 the mound was partially explored by W. M. K. Warren who dug a trench through the middle of it. He reported 'In the centre we found an old decayed oak beam which we followed down to a depth below the level of the field. We found it fixed to an octagonal oak frame, which we considered to be the ground frame of a windmill, whose foundation had been a mound placed on the top of this frame'.[26] The mill was evidently a sunk post mill, for a tower mill would not have been carried away as described, but the ground frame was apparently unconventional, if Mr Warren's description may be relied upon. A windmill is shown at Chedzoy on Andrew Paschall's unpublished map of the Sedgemoor Battlefield of 1685.

CHEW MAGNA *approx. 571 658*

A field name suggests the former site of a windmill on the south side of Dundry Hill as recorded in the Tithe Award.

CHILTON POLDEN *approx. 371 400*

A windmill site is suggested by a field name in the Chilton-upon-Polden Tithe Award.

CHURCHILL *approx. 445 601*

There is a Windmill Hill here and Knight says 'It may be observed in passing that the mill which gave the eminence its name was standing there at least as late as 1652'.[27]

COMPTON DUNDON *1. approx. 495 339*
2. approx. 496 319

A windmill is marked on the hill still known as Windmill Hill by Newcourt, c. 1660, and Ogilby in 1675, and the hill is now marked by the monument erected to the memory of Admiral Hood who was born at nearby Butleigh. The second site, about a mile and a quarter to the south, is suggested by the name 'Mill Way' in the Tithe Award.

COSSINGTON *approx. 356 400*

John Ogilby's road map indicates 'Cossington windmill' at this site in 1675.

CURRY RIVEL *388 246*

Willtown windmill is marked on Greenwood's map of 1822 and worked, from information obtained locally, until some time between 1874 and 1884.

ASHTON MILL, CHAPEL ALLERTON

The man at the mill door is Tom Petheram, so this photograph was taken before 1897. Of interest is the whitewashed tower and broad, common sails, one pair clothed each side of the whip. The weathercock was a typical Somerset tower mill feature.

Ashton Mill at work with a pair of common and a pair of spring sails after being refitted in about 1900. The steps to the door at the side of the tower have gone, the thatched cap rebuilt and the tower, still white-washed, is bound with three iron hoops. To the right of the picture a masonry loading stage can be seen on the edge of the mound. The miller is John Stevens, the last windmiller in Somerset.

Above: the brake wheel and iron wallower, superseding the wooden one, in Ashton Mill. Two of the wheels which centre the cap on the curb can be seen behind.

Left: The single pair of stones in Ashton Mill, with an iron crane built in close by the upright shaft to the left, to raise the runner stone for dressing.

The horse supporting the hopper over the millstones, with the shoe beneath. A spare shoe rests on the stone casing. The small distance between the upright shaft, right, and the tower wall behind shows how little room there was in Somerset tower mills.

The guarded great spur wheel on the bottom of the upright shaft in Ashton Mill, with the sack hoist chain in the foreground, and bridge posts and bridge tree behind.

The tentering for the single pair of stones in Ashton Mill, with the drive to the governor from a pulley on the stone spindle. The tower wall has been notched to take the swing of the governor when the mill was at work.

The crown wheel and sack hoist drive in Ashton Mill, with the dog clutch shaft coupling just below the crown wheel on the upright shaft. The lever to the left lowers the pinion into gear.

It was apparently one of only two mills built in the county with a stage around the tower, and the tower must have been of very narrow diameter at the curb, for it is only about 12 feet in diameter at ground level and had a marked batter. Remains of the tower to a height of about 9 feet exist, built of neatly squared lias blocks, and the tower was brought down to this level by about 1930. Nothing is known of the internal layout of Willtown Mill, but it apparently had quite a short working life. The last miller, certainly the last tenant, was John Solway.

CURRY RIVEL, Wick *403 265*

Bowen's map of 1760 shows a windmill on Wick or Week Hill to the west of Langport, and an estate map of Week, dated 1768, shows a windmill in elevation with the annotation ' Wind Mill new Built '.[28] This appears to be a post mill, possibly with the substructure cased in solid masonry. In 1775 a ' New erected Wind-Mill on Week-Hill, near Langport ' was advertised to let, and said to be ' of an excellent construction, consisting of three pair of stones with every conveniency annexed, and is now in perfect order '.[29] Applications were to be made to Maurice Trevillian Esq., whose name also appears on the earlier estate map. It would seem unlikely that the windmill on the map of 1768 could have contained three pairs of stones, but it is a matter for conjecture whether map and newspaper advertisement refer to the same mill. On an estate map of Drayton with Curry Rivel of 1820 the site is marked by ' Windmill Toft ',[30] but nothing apparently remains at this location now.

DITCHEAT *approx. 627 341*

' Blindmill ' is recorded at this location in the Tithe Award, the site being on Acres Hill, between the rivers Alham and Brue.

DURSTON *1. 294 280*
2. 297 278

Two possible sunk post mill sites exist at these references. Cross walling was discovered in the first mound when a reservoir was built in it in 1962, and the second site is marked ' Mill Mound ' on the 1″ OS map.[31]

EAST BRENT *356 518*

A windmill is marked on Ogilby's road map of 1675 and on Bowen's county map of 1760, and on the early 19th-century OS maps and Greenwood's map of 1822. The site is apparently that of a post mill which stood on a mound, which is still visible, and worked perhaps until about 1880. The farm is still known as Mill Batch Farm and the farmer related that he had discovered a stone causeway running roughly parallel with the road between the house and the mill mound, although not necessarily connecting them. He could not recollect seeing any debris when ploughing over the mound. Windows of the former farmhouse bore scratched signatures, latin text and other matter and dates between 1824 and 1860.

EAST BRENT, Edingworth *357 530*

There was a post mill here in a field still known as 'Mill Moot', which was described as a 'private mill'. This expression, as used hereabouts, apparently meant a mill used on a large holding mainly for grinding the holder's own produce. The mound was removed by a Mr Gillian about 1914, 'despite the protests of his neighbours', so here antagonism to destruction was rooted remarkably early in terms of windmills.

EAST QUANTOXHEAD *1. 119 432*
2. 128 433

Two possible windmill mounds exist on the north Somerset coast at the head of the Quantock Hills.[32] The first mound is in a field marked as 'Windmill Piece' in the Tithe Award.

EVERCREECH *651 387*

A mound east of the Bell Inn is possibly the site of a former windmill.[33]

GLASTONBURY *1. approx. 519 390*
2. approx. 505 395
3. approx. 496 384

'Molendina ventorum/molendina ad ventum' are recorded at Glastonbury in circa 1244 and 1309, and a windmill referred to at 'Stonyacre' in 1305-6 was perhaps on the present Stone Down Hill.[34] Collinson quotes a terrier of the penultimate Abbot of Glastonbury, Richard Beere, (1493-1524), which refers to a 'water-mill in the town and a wind-mill above it', possibly on the present Windmill Hill.[35] A windmill is marked on Wearyall Hill on Newcourt's map of 1660, Ogilby's road map of 1675 and this mill was set up by one John Wrinckmore 'for the grinding of wheat, malt, pill corne, and other graine, which mill hath grown (sic) corn for the most part of the town ever sychence. It is a great hindrance to Beckery Mills for many of the Customers which heretofore did grind at Beckery Mills do now usualyy grind at the said new-erected windmill'. John Wrinckmore was tenant of the Crowne and a horse-mill belonging to it in 1640, and travelled to Ireland for eighteen months to escape arrest for debt. He was clearly unpopular in the town, even more so for setting up a windmill in opposition to the town watermill.[36] A pair of engravings of c.1775 show a windmill on Wearyall Hill in the background of the Abbey ruins,[37] but neither Bowen's map of 1760 nor Day & Masters' of 1782 show a windmill near Glastonbury.

GLASTONBURY, Common Moor

The draining of Common Moor was completed by 1722, the first Parliamentary Enclosure Act in Somerset, and some form of water-pumping engine, possibly wind-powered, was used.[38] In the diary of physician Dr Claver Morris an entry notes 'we went & saw Mr Nicholl's engine to throw the water into all the Ditches of Common-Moor, & he set it a working on purpose for us '.[39]

GREINTON *approx. 417 371*

A windmill is marked near the Polden ridge above Greinton on Newcourt's map of 1660 and Ogilby's road map of 1675. It had gone by 1742 when the name 'Windmill Piece' appears on an estate map.[40]

HARDINGTON MANDEVILLE *approx. 511 114*
approx. 521 126

A Windmill Hill exists, and a mound was visible in about 1930 at the first reference. At the second reference, towards the parish boundary with West Coker, a Windmill Cottage is marked on the modern 6″ OS map on Hardington Moor.

HENSTRIDGE *1. approx. 706 213*
2. approx. 700 198

A Windmill Hill still exists at Yenston and on Toomer Hill, further to the south, 'Windmill Field' is marked in the Tithe Award.

HIGH HAM *1. approx. 435 307*
2. approx. 413 317

Newcourt's map of 1660 shows windmills at Sedgemoor Hill and Turn Hill and Bowen's map of 1760 indicates a mill at the former site, which is some 200 yards north east of Stembridge Mill. This mill is drawn in elevation as a tower mill on an estate map of the Manor of Netherham of 1779[41] and probably worked until the early 19th century when Stembridge Mill was built to replace it.

HIGH HAM, Low Ham *1. approx. 433 285*
2. approx. 427 298

A windmill is marked at Low Ham on Newcourt's map and the estate map of 1779 contains a drawing of a derelict tower with the annotation 'Windmill in decay'.[42] The site appears to be marked by a break in the line of a hedge. The second site at Low Ham, about half a mile to the north west of the first, is marked 'Mill Mote' on the estate map, possibly a windmill site or a field connected with the nearby Paradise watermill.

HIGH HAM, Stembridge Mill *433 305*

Stembridge Mill was built in 1822 on what was part of the Eastfield, enclosed in 1792, and incorporated some gear from the tower mill known as Ham Mill which stood nearby. The tower, of random coursed lias stone, is parallel-sided to above the heads of the first floor opposite doorways and then tapers quite steeply to the curb. The thatched gable-shaped cap is now unique in England, although many windmills once had thatched caps, and the mill was winded by an endless chain geared to a worm meshing with a rack on the outside of the curb. In 1897-8 the cap jammed, apparently due to distortion of the curb, and the mill ceased to work by wind.[43] The sails have no stocks, the frames being supported by backs which are bolted to an iron coffin cross fixed into the end of the timber windshaft. The sails were set from a circular earth and masonry mound built up around the base of the tower, and access is made to the tower at lower ground level on the lee side through the mound and to the opposite first floor doorways from the mound level. The mill contains four floors and incomplete gear which once drove two pairs of stones. An interesting feature which occurs in several Somerset tower mills is a fireplace, located in the wall of the lowest floor, with a flue running up within the thickness of the tower wall to an outlet just below the cap. The brake wheel is of clasp arm construction and was probably salvaged from the earlier mill, bearing notches which indicate it was once supported by compass arms, and has morticed wooden cogs. The over-drive to the stones, an upright shaft which carried a spur wheel and the pinions and spindles it drove, is missing, but the two pairs of 4′ 0″ diameter stones are in place. Those at the front of the tower are French Burr, the runner carrying balance weight pockets patented by Clarke & Dunham in 1859 in its plaster back, and those at the rear are of sandstone conglomerate bedstone and French Burr runner stone and were latterly under-driven by bevel gearing from a portable steam engine set up outside the tower. This gear is still in place, and the horizontal shaft from the engine drive also carries a belt pulley which drove the wire machine which is situated behind the mill trough. There were no bins built into the mill and the hoppers and asociated timberwork as well as the cases enclosing the stones have gone.

John Sherrin was miller at Stembridge Mill in 1823 and in 1824 he inherited the mill and cottage from his father-in-law. His three sons, John, Robert and Henry inherited the property at his death and Robert is recorded as miller in 1861 and 1866. He was probably succeeded by Simon Spearing in 1869 and in 1879-81 Joseph Loader was working the mill, but the length of his occupation is uncertain. George Parker was renting the property in 1889 and at this time the first reference to steam power being used also occurs. Frank Parker succeeded his father in 1897 but moved away a year later and the next occupant was Robert Mead. It was during Mead's occupancy that the bakehouse was last used by James Duddridge. The bakehouse building can still be seen close beside the mill tower. By 1906 Robert Hook had bought the property, but sold it to Dr Bellot by 1910. The last miller's name has been given as F. G. Harding and as Mr Hill, and both may have been millers during the last years of working by steam. The mill passed to Dr Bellot's son in 1928 and in 1969 the property was acquired by the National Trust and partly restored by them.

The mill house which stands nearby has the name 'Simon Spearing' and the date 12 November 1869, also the names 'F. Parker' and 'G. Parker' scratched on a ground floor window pane and the mill also carries the names of former millers and workmen, amongst them F. W. Lavis, wheelwright of Long Street, Ernest Fisher, thatcher of Pitney Hill, and J. Spearing with the date 1898 written in red oxide on a stone floor beam.[44] The wheelwright and the thatcher were probably among the many rural craftsmen whose work is represented, usually anonymously, in Somerset windmills.

HUISH EPISCOPI *approx. 415 286*

Bowden's Mill stood on a knoll overlooking the main Aller-Langport road, above Combe. It was a stone tower mill and worked up to about 1874; the tower stood until about 1885.

HUISH EPISCOPI, Pibsbury *433 265*

Pibsbury is about two miles out of Langport on the Long Sutton road and is bounded by a rectangle of roads. The windmill stood at the north east angle of the lesser road and was built by William Wheller between 1797 and 1823, costing him 'many hundreds of pounds'.[45] Pibsbury Mill was the tallest known Somerset windmill, and formed part of a group which incorporated the miller's house, barn, cart-sheds and stables, with facilities for loading and unloading. The tower was parallel-sided up to a timber stage, from which the sails were set, and then tapered towards the curb. The cap was thatched and the mill was winded by hand-operated luffing gear in the tail-box. For the last twenty years of its working life, however, the cap was covered with wood and felt, the thatch having blown off in a gale. Four common sails drove, via wooden gearing, two pairs of 4′ 4″ stones, one pair of French Burrs for flour and one pair of sandstone conglomerate bedstone and French Burr runner stone for gristing, as at Stembridge Mill. The mill worked until shortly after 1897 and part of the tower collapsed on a stormy night in April 1915.[46] Some of the gear was bought by Simon Spearing of Stembridge Mill and in the early 1930s some was still in use at Isle Brewers watermill, worked by his son. In 1921 much of the tower was removed and the stone was sold and re-used to build a new house nearby. In the early 1930s remains could be found on site, including a millstone which had been reduced from 52″ to 46″ diameter for re-use in a watermill at a cost of 2/-, but was never removed. The site is now difficult to trace.

John Hookway was the miller at Pibsbury in 1861 and William Wheller is recorded as miller and beer retailer there in 1872 and ran the windmill to the end of its working days. He also ran the local watermill and in the 1880s is recorded as miller, beer retailer and saw mills, using wind, water and steam power.

An interesting description of Pibsbury Mill is given in Walter Raymond's *In the Smoke of War*,[47] a story set in the Civil War some 150 years before the mill was built, but Raymond evidently knew the mill well, although the details of the inside of the mill are too inter-woven with the plot to piece together an accurate picture.

HUISH EPISCOPI, Wearne *approx. 424 298*

James Courtenay owned a windmill in Wearne in 1585 and sailcloth was stolen from Col. Stawell's windmill at Wearne in 1664. 'Milltoyt Way' occurs in 1620, referring either to this site or that of Bowden's Mill,[48] and a windmill is marked on Newcourt's map of 1660. 'Windmill Hill' is marked as a field name on the Netherham estate map of 1779, and the mill had apparently gone by that time. A mound is still visible in this area.

HUTTON *361 589*

In 1309 an Extent of the Manor of Hutton mentions a windmill with an estimated annual value of 26s 8d, and a detailed partition of the manor of 1430 makes no reference to a windmill but refers to a plot of land near 'le Milstede'. A rental of 1482 of the estates of John Payn of Hutton included one windmill 'de novo' edification with an annual value of 20s.[49] Hutton Mill is marked on the 1st edition OS map of 1809 and William Hookway was occupier and part proprietor in 1817.[50] It was he who built the house, the present windmill cottage, and enclosed part of the hill as a garden. Samuel Sarjeant appears as owner and occupier in 1844 and in November of that year advertised the mill for sale as a lucrative milling and baking business.[51] A Mr Nichols was miller in about 1848, followed by Thomas May until about 1853. From 1854-6 Henry Hiscock is recorded as miller, followed by Thomas Follott, Joseph Wilson and Thomas Carnell. Each of these men were apparently at the mill only for a short time, for by 1861 Alfred Bowering was miller, and it was probably he who worked the mill last in about 1864.

By the early 1930s the tower of Hutton Mill had almost collapsed but was subsequently much rebuilt and used as a Royal Observer Corps post during the 1939-45 war. The tower still stands to a height of about 18 feet and is used as a garden store; fine views over the Mendip Hills, Weston Bay and Worle Moor are afforded from the top of the tower.

HUTTON, Oldmixon *approx. 337 580*

John Strachey's map of 1736 and Bowen's county map of 1760 indicate a windmill at the west end of Bleadon Hill and an Oldmixon estate map of 1756 made by John Hine shows a windmill on the open hillside just within the Hutton parish boundary.[52] It does not occur in the parish records, however, and there are no records outside these dates. The supposed site consisted in the early 1950s of a low circular mound suggestive of a tower mill, but the dating and the effect on the site of ploughing make it difficult to draw any definite conclusions.

KENN *411 696*

The tower of Kenn Mill, now much shortened, carries a stone carved with the initials 'C.B.'[53] and dated 1821, and this windmill was the first to occupy this low-lying site on the north of Kenn Moor. The mill was a three storey tower mill with a single pair of stones and the thatched cap is said to have been winded by a tailpole reaching to near the ground, a method rare if not unique in the county and perhaps surprising on a windmill built in the 19th century. The mill worked until sometime towards the end of the 19th century, but was powered also by an oil engine from the 1880s and it is possible that its low position on the Moor, only some 20 feet above sea level, accounted for this. The last miller was a Mr Herbert.

The tower was truncated in about 1900, the wall formed into a gable shape and the top thatched, and before the great war it was used as a slaughter house. In the 1930s the millstones, in pieces, remained at the farm and during the 1939-45 war the tower was used as a Home Guard lookout. The tower has now been reduced to an uneven height of about 18 feet and is completely gutted owing to the fact that it was used recently as an incinerator. The ground floor contains a fireplace, with the flue within the tower wall, and timber from the machinery is said to have formed the raw material for Clevedon War Memorial.

There were dim memories here of a happening possibly linked with this mill. In 1830 three men were sentenced to death for setting fire to some mows of wheat here and were ordered by the High Sheriff to be executed on the spot where the crime was committed. Fourteen thousand persons collected on Kenn Moor and a troop of cavalry and 200 constables were in attendance. One of those executed, named Wall, left a wife and seven children.

KINGSDON *approx. 512 269*

In 1628 a windmill formed part of the estate held by Thomas Browning and a windmill was mentioned in the North field in 1694.[54] A Mill Lane still exists and site of the mill was probably at the north west end of the lane on Kingsdon Hill, from field names in the Tithe Award.

LANGPORT *approx. 422 268*

A windmill is first mentioned at Langport in 1344, known as the 'Big Mill', and was farmed out by the lord in 1349-50 for 26s 8d. It was blown down by a 'great wind' in January 1362 and was evidently never rebuilt, the name being perpetuated in a mound known as Windmill Toyt.[55]

LOCKING, Vale Mill *359 617*

Vale Mill was built about 1813 and worked by wind alone until 1889, when a steam engine was bought from a Mr Harris, who had used it for bean crushing, and hauled by horses to the mill. It proved unsatisfactory, however, and was re-purchased by its original owner. W. Long was miller in 1861 and Thomas Quick from about 1866 until about 1880 when his son, Thomas Quinten Quick, took over and worked the mill until sometime between 1906 and 1910. Mr Quick died in 1916 and the sails were removed from the mill in about 1922 and stored. The mill remained intact and in good order until

July 1962 when it was vandalised and gutted by fire, and the remains of the masonry shell have since been incorporated into a house.

Vale Mill was a typical Somerset tower mill, a parallel-sided stone tower with a gable-shaped cap, originally thatched but later covered with corrugated iron. The cap was winded from inside, the gear being housed in a tail-box, and the four common sails drove two pairs of stones, through spur gearing on a pitch pine upright shaft. The gear was predominantly of timber, even though the mill was of 19th century origin, a clasp arm timber brakewheel mounted on the timber windshaft and a clasp arm timber spur wheel on the upright shaft which drove the two pairs of stones from below. Like Ashton Mill, Vale Mill was fitted with a governor to regulate the distance between the stones when milling, and the governor was driven by belt from the upright shaft below the stones. There were no big bins in the mill, the corn being emptied into the hoppers of the stones as required in the usual Somerset windmill fashion and the dressing machinery and sack hoist were driven from a crown wheel on the upright shaft on the second floor.

When the windmill was sold, intact except for the sails, in 1960 for £850 several milling firms had expressed an interest in preserving the property and it was said that, with new sails, the mill could have been put to work by wind again within ten days. It was therefore a great loss to the county when the mill was so completely destroyed by fire only two years later.

LONG ASHTON

A mediaeval windmill was attached to the estate of the family de Lions in Long Ashton, for in about 1312 William de Lions held, at his death, 'a capital messuage . . . a hundred and forty-seven acres of arable land, forty four acres of meadow, and separate parcels of pasture, with a windmill, two fulling-mills and diverse other possessions'.[56]

LONG SUTTON, Upton *approx. 454 266*

A windmill in Long Sutton was held by Athelney Abbey in 1349 and was let for 10s. in 1538-9.[57] It may have stood in a field to the north west of the parish which is marked as 'Windmill Toit' in the Tithe Award.

LYMPSHAM *326 551*

A plan of Batch Farm, Lympsham, 'taken by sight on horse back without scale or compass by Calib Dickinson as the same was shown to him by John and Thomas Story two of the Tennants 18th and 19th Sept. 1758' shows in elevation a tower mill on a mound.[58] The mill had apparently gone by the 1830s and is not recorded on any of the county maps.

MARK *387 378*

In 1700 a Wedmore surgeon, John Westover, disestablished his windmill at Wedmore and 'lent cozen William Westover my crane, fower pooley wheals one toung clavey and two toung clavey pinns, a payer of crewes and plainke for ye screwes. Lent him before 18 veallows to bare ye mill'.[59] The Reverend Harvey, who transcribed John Westover's diary, considered that the surgeon had 'sent the sails etc.' to his cousin's mill at Mark, but 18 men and the lifting tackle involved rather imply the re-erection of a windmill. A mound can be seen in a field opposite Mill Batch Farm on the road from Wedmore to Mark, perhaps the site of Westover's Mill. There is an account in the Mark Overseers' Book of a 'Great Gale' of 26 November 1703 'blowing down houses and mills' and also an account of a gale which uprooted two mills 'at Mark Causeway', which is between Mark and Watchfield, so possibly there were more windmills at some time in the parish.

MEARE *approx. 451 416*

'Mill Batch Fields' are locatable at this site from the Tithe Award.

MIDDLEZOY *approx. 366 328*

A windmill is marked on the little island of high ground west of Middlezoy village on Newcourt's map of 1660 and on Greenwood's map of 1822. The last mill stood about a furlong or so north of Thorngrove and in about 1930 the remains of the foundation of a stone tower mill could be seen. The mill evidently worked with a single pair of stones and the gear was originally of timber, but replaced by iron before the mill ceased work. Dennis Denman was miller at Middlezoy until he left to work at Woolavington and the last miller was Ira Meaker, who was at the mill in 1872 and the mill ceased work about 1885.

MONTACUTE *approx. 506 179*

A windmill belonging to Montacute Manor was held by Thomas Phelips in 1560 and three fields retained windmill names in about 1700.[60] Windmill Farm still exists and is marked on the current OS maps.

MOORLINCH *400 372*

A windmill is marked on Knoll Hill, a fine lofty outpost above the village of Moorlinch on Newcourt's map of 1660 and Ogilby's road map of 1675, also on Bowen's map of 1760. According to a local report the tower mill which was the last occupant of this site was built to replace a post mill,[61] but the date of building is unknown. A windmill was standing early in the 19th century and a watercolour of c.1840 almost certainly shows the tower mill.[62] The mill worked until about 1889 and was soon afterwards pulled down and the masonry used to build a low shed which can still be seen. The mill-stones, some shafting and gearing, wood pulleys and the entire luffing gear were bought by John Stevens, who had learned his milling there, and installed in Ashton Mill, Chapel Allerton. The land was bought by a Mr George Pitman. It is said locally that in certain lights the steps which led into the mill can still be discerned, but the site is now much overgrown. Joseph Loader, the brother-in-law of John Stevens, was miller at Moorlinch in 1883, Frank Spearing worked at the mill before going to Watchfield Mill and the last miller was Dennis Denman, originally from Middlezoy and Woolavington windmills.

NORTH CHERITON *approx. 686 258*

A windmill at North Cheriton is indicated by an annotation on a map of a Survey of the Manor of Latterford of 1766.[63]

NORTH CURRY *approx. 315 248*

'House and Windmill' are recorded in the Tithe Award of c.1840 and a Windmill Hill still exists.

OTHERY *approx. 377 313*

A windmill is marked on this low island of ground, similar to Middlezoy close by, on Newcourt's map of 1660 and 'Mill Batch' is marked on the large scale OS maps. The site has now been built over by a garage.

PAULTON *656 567*

A mound some 22 paces by 3′ 6″ high with a wide and shallow dip in the top survives in a field called Windmill Ground and is probably one of the few windmill sites to have existed in the north east of the county.[64]

PAWLETT *1. approx. 291 442*
2. approx. 293 432

'Mill Moot' is recorded in the Tithe Award east of the present Yearsey Farm and 'Mill Batch' is marked about a mile to the south on Pawlett Hill. Local memory recalled a 'well-known mill', but nothing further is known.

PAWLETT, Hams *277 411*

This was never a windmill in the conventional sense although recorded by the usual windmill symbol on the 6th edition 1″ OS map. The site, at the far south Pawlett Hams on the north bank of the River Parrett, was formerly occupied by a wind pump which was used to bring fresh water from the Cannington Brook opposite under the tidal Parrett. The appearance of the pump at a distance was evidently that of a brick tower with four sails and a head or cap luffed by a trailing chain. Wind power was superseded by steam in the 1880s[65] and when the pump was pulled down the bricks were re-used to build a culvert. Pumping was latterly done by a boosted siphon. Why the symbol of a windmill appeared on maps years after the wind pump had gone is a mystery.

PITNEY

In 1605 Edward Clawsey was said to have recently erected a windmill and in 1691 the manor of Pitney Lorty included a watermill and a windmill. A long narrow field terminating in the south west of the parish is known as Mill Close.[66]

PORTBURY *approx. 509 740*

A windmill is marked on Windmill Hill on Donn's map of 1769 and 'Okey Mill' is marked on Day & Masters' map of 1782, but curiously so, for there is no windmill symbol and the other sites marked on this map of Somerset have only symbols with no annotation. Day & Masters' site is close to the now ruined court, whereas field names in the Tithe Award suggest a site further east.

PORTISHEAD *458 767*

An indenture made on 25 March 1832 between the Mayor, Burgesses and Commonalty of the City of Bristol and John Nesbitt of Portishead, miller, records that the stone tower mill and cottage were recently built or building and were to be 'completed according to a plan and elevation already submitted by the said John Nesbitt to the Surveyors of City Lands'. The lease was for 99 years at a rental of £3 per annum. By 1846, however, it appears that John Nesbitt had left the mill and all attempts to let it again failed. A legal letter of 4 May 1848 refers to the mill having 'been unproductive for nearly two years past' and suggests that the tower be converted into a dwelling to provide additional accommodation for the tenant of the mill cottage, 'in consequence of there being a Mill in Portishead worked by Steam, there was little prospect of the Mill under the lease being worked to any advantage'. Permission was granted for the conversion on the understanding that no new building was to be permitted on the land 'without the elevation—being first submitted to the Committee', and a Mr Gay put in an estimate to carry out the conversion in 1848. The proposals included removal of the machinery, putting a new roof on, new floors, a staircase and a chimney stack and the work was apparently undertaken as planned.[67] There is a story, set between 1860 and 1865, which refers to the windmill being used by smugglers[68] and a watercolour apparently dating from about 1875 showed the mill with a thatched head,[69] but it would seem likely that memories of the mill lived longer than the mill itself, for Portishead windmill appears to have been one of the last built in Somerset, to have had one of the shortest working lives and to have formed one of the first windmill house conversions, a sad tale. The site is very close to the sea and the mill gave its name to Windmill Bay just below, and it is stated that the mill never did much work because the situation was too low and 'people brought their corn there and had to wait sometimes six weeks before the wind was suitable'. The straight tapered stone tower still stands, incorporated in 1908 into a building for a golf club house, which is now a public house.

PURITON *1. approx. 319 411*
2. approx. 325 409

A windmill is shown at Puriton on Speed's map of 1610, and on Ogilby's road map of 1675 two windmills are shown, annotated 'Purito Windmills'. Only one mill appears on Bowen's map of 1760 and there is some confusion because the two sites are close together and descriptions of the exact locations are not always specific. The first site, however, is just north of the footpath leading from the Puriton-Knowle road to Dunball station and in about 1930 this footpath was flanked by some acres of cultivated teazles just ripe for harvesting, a sight which is not likely to be seen again. At this time remains of a mound and the circular stone foundation of a tower could also be seen and a Windmill Cottage still exists. The tower mill last worked some time between 1870 and 1875, the last miller being Joe Crane from at least as early as 1866, and was pulled down in the early years of this century when the stone was re-used to build 'some sort of tower' at Puriton Manor.

The second site, to the east of the first, is marked 'The Windmill' in the Tithe Award of about 1840, and this coincides with the windmill marked on the 1″ OS map of 1817. 'Windmill Furlong' is also marked in the Tithe Award to the north west of this site, possibly the site of another, earlier, windmill or connected with one or other of the known sites. Remains at the second site were not found.

PUXTON *approx. 393 636*

'Mill Moot' occurs in the Tithe Award.

ST. CUTHBERT OUT, West Horrington *approx. 577 474*

'Mills Field' occurs in the Tithe Award, the field being to the east side of the road running between West and East Horrington. A slight mound can be seen in the south west corner of this field and this site is the only one known on the Mendip scarp.

SANDFORD ORCAS *approx. 625 218*

A Windmill Hill exists in this former Somerset parish, now part of Dorset.

SEAVINGTON ST. MICHAEL

The earliest-known reference to a windmill in Somerset, c.1212, as quoted earlier, refers to this parish.[70] No later references are known.

SHAPWICK *1. 422 373*
2. 425 374

A windmill at Shapwick, part of the Glastonbury Abbey estates, was valued at £3.6s.8d in the early 1330s.[71] Newcourt's map shows a windmill here in 1660, also Bowen in 1760 and Day & Masters' in 1782. Again there is a possibility of confusion between two known sites, one each side of the road, and information tends to appear inconsistent.

Shapwick post mill stood in the north west angle formed by the Shapwick to Pedwell and Bawdrip to Glastonbury roads and was blown down in a gale on the night of Tuesday 29 November 1836. The miller, William Jones, was so injured that he died ' in a few hours ' and his brother, who was working the mill with him, was rendered unconscious. At the inquest on William Jones it was remarked that another post mill ' in an adjoining parish ' had been blown down twice ' and we have heard of several others having met a similar fate '.[72] The windmill is referred to at this time as a ' stump mill ' of timber construction, suggestive of a post mill with its substructure buried in the ground. This is borne out by the fact that in 1934 the ' footprint ' of this mill, a cross shape on the mill mound, could be seen during the dry summer.

Shapwick tower mill was sited about a third of a mile from the post mill, near the eastern parish boundary. The date of building is unknown, and the maps are uneasy evidence for the windmill location varies from side to side of the road depending on the cartographer. Little information was available locally, for this mill evidently ceased work earlier than most, perhaps because of the steam mill nearby. It is said to have worked until about 1860 and, if built to replace the post mill, would have had a working life of only about 24 years. If it was built to replace the post mill the reason for choosing a different site was almost certainly because it was prohibited to build a windmill closer than 200 yards from a turnpike road. This may also explain similar siting differences, as at Watchfield. The remains of the foundation of the tower, built into a small mound, can still be found and the mill is said to have been a typical Somerset tower with a thatched cap.

SOMERTON

There were several windmills recorded in and around Somerton, the earliest reference being to a windmill ' de novo levatum ' which was part of Somerton Manor by 1275-6 and was still in existence in 1330. By 1334 there was a windmill also at Mileburgh, now Melbury (approx. 480 275) which had gone out of use by 1484. A grain mill which was sold in 1575 was described as a windmill in 1619 and stood in South field. It may have been the ' old mill near St Cleers Pitts ' mentioned in 1715 and probably stood in a field called Mill Close, later cut by a road and the railway. It was known as Applin's Mill and had evidently gone out of use by 1802. It was evidently in use throughout the 18th century and the maps of Strachey, 1736, and Bowen, 1760, show a windmill at Somerton (approx. 477 283). There was a windmill on Crane's Farm by 1616 which retained the name Crane's Mill into the 18th century, was working in 1779 but had gone by 1802. The Tithe Award shows ' Crane's Mill ' at approx. 490 275. Another windmill in Somerton parish was devised by William Champion of Shapwick in 1650-1, although its location is unknown, and a windmill on South Hill, marked on Ogilby's map of 1675, was owned and occupied by John Coolin in 1685. (approx. 471 271). It was known as Cullen's Mill in 1692 and was still standing in 1749.[73]

STAWELL *1. approx. 368 389*
2. approx. 376 386

Newcourt's map shows a windmill at Stawell, above the village on the Polden ridge and Ogilby marks two in 1675, the first on the present Stawell-Bawdrip parish boundary, qualified by field names in the Tithe Award, and the second shown as 'Stoll Windmill', on Cock Hill. Local inhabitants in the 1930s had heard a tradition of a windmill here, referred to as 'Cock Hill Mill', the site south of the road near the so-called Chilton Priory.

STAWELL, Sutton Mallet *369 365*

Andrew Paschall's unpublished map of the Sedgemoor Battlefield, 1685, shows a windmill, Sutton Mill, south west of Sutton Mallet church and the site is marked by 'Mill Batch' on the larger scale OS maps and by a mound which survives on Sutton Hams.

STOCKLAND BRISTOL *249 439*

Part of the living of the vicarage, instituted in 1317, included the tithes of two windmills[74] and 'Old Mill Mound' is recorded on the 1″ current OS map,

STOKE ST. GREGORY *1. 339 277*
2. 354 269
3. approx. 364 281

On Windmill Hill, between the River Tone and the road to North Curry, a mound with a depression in the top still exists. In Park Meads, Woodhill, a brick Windmill Cottage exists by the site of the 'Windmill on Woodhill Green' referred to, and illustrated as a typical Somerset tower mill, on the Rev. William White's Map of the Manor of East Curry, otherwise Stoke St. Gregory, of 1787. In the reference book accompanying the map the tenant is recorded as the Widow Dawe.[75] The windmill is mentioned in the Parish Book in 1704,[76] but does not appear on the 19th century maps. The third site, to the north east near Stathe, is suggested by field names given in the Tithe Award.

STOKE TRISTER *approx. 736 286*

A field south west of Stoke Trister church is called 'Windmill Field' in the Tithe Award.

STREET *1. approx. 492 362*
2. approx. 477 345

A windmill at Street belonging to Glastonbury Abbey is recorded in c.1255.[77] The maps of Newcourt, 1660, and Ogilby, 1675, show a windmill east of the town in Millfield, where Greenwood's map of 1822 records 'Mill Batch', and also a windmill at Ivythorn, to the south. Ogilby calls this 'Ivythorn Windmill'. The former site is now well-known for the famous school called after it.

UPHILL *317 583*

This building illustrates the difficulties of the windmill historian. Situated near a town, outstandingly visible, on much frequented ground, a landmark for ships and in a parish remarkable for the longevity of its inhabitants, local enquiries only served to cast doubt that this ever was a windmill, the official view being that it was built as a watch tower. Artillery was mounted on this hill against the coming of the Spanish Armada and a map of the time of Henry VIII shows a block-house standing near the old Uphill church.[78] Day & Masters' map of 1782 shows a tower only on the hill, but a Royal Exchange Fire Insurance Policy of 11 August 1789 gives positive evidence of a windmill : ' John Marchant of Uphill in the Co. of Somerset, Yeoman. On a Windmill situated on an eminence near the Church in Uphill aforesaid and known by the name of Uphill Mill, £100 '.[79] By 1829, however, it was referred to as ' the walls of a windmill '[80] and on 10 March 1863 a beacon was built on iron rods laid across the top of the tower to commemorate the wedding of the future King Edward VII. A lad, one Oliver Counsell, ' borne up a ladder reared against this side of the pile ' set it alight. Corn is said to have once been grown on both sides of this hill as well as on the northern slopes of neighbouring Brean Down. This would probably have been during the Napoleonic Wars and in all probability the end of these wars saw the end of this mill and of corn growing on this sort of ground. The tower, rebuilt with a castellated top and an iron spiral stair for use as an observatory in 1934, still stands on a fine, but very exposed, vantage point.

WALTON *462 352*

An account of the building of a post mill at Walton in 1342-3 has already been mentioned[81] and while there is no evidence, for or against, that this mill occupied the site of the last Walton Mill, the tower of which still stands, there is evidence of occupation of this fine site from the middle of the 17th century, when a windmill is marked on Walton Hill on Newcourt's map. Ogilby also shows ' Walton Windmill ' in 1675 and an indenture of 1725 refers to a ' Wind Grist Mill lately standing . . . but now fallen down '. By 1741 a new-erected windmill, referred to as a ' stone-windmill', had been built by John Gipson of Ashcott, a miller, and was leased by him to Peter Ralls. In 1793 the mill was purchased by Thomas, Marquis of Bath and Thomas Davis of Longleat for £25.5s.0d. and subsequently rebuilt by them, for in 1797 James Chancellor of Walton was granted a lease of a ' new erected Windmill sit on Walton Hill with a plot of Ground adjoining ' by the Marquis. In 1826 the mill was leased by James Chancellor, son of James, Samuel Wilkins and his wife, Mary, also with the house, garden and bakehouse.[82] William Crane was miller from at least 1861 to his death in 1889 and the last miller was Charles Phillips, who ran the mill with a bakery and grocery business, until about 1906. The mill was bought by the Rev. G. M. Evans of Westonzoyland and in 1926 was converted into a summer residence by him, the tower being heightened at this time to form a parapet and look-out platform, with impressive views over King's Sedge Moor.

Walton Mill was a typical Somerset tower mill, although the tower is of slightly greater diameter than most others surviving. The mill had four common sails, a thatched gable cap with winding gear housed in a tail-box, and drove two pairs of stones. The gear was largely of oak and elm, with apple wood cogs. Three French Burr millstones are embedded in the ground outside the tower.

Frances Hariott Wood in her story *The Brothers* tells of twin sons who could only be distinguished by their father ' and the maid up at Walton Mill, the miller's only child who was in love with both young men, as they were wildly in love with her '. One brother fought for Monmouth at Sedgemoor and the other, going to his rescue after the battle, was captured in mistake for him. Before being taken he had told the fugitive to make for the mill and hide, which he did, but Bina, the miller's daughter, reproached him so bitterly for deserting his brother that he ran away. The captured brother was hanged on Marshall's Elm as one of Monmouth's men.[83] This story is based upon known facts, but on an earlier Walton Mill, for Sedgemoor was fought in 1685.

WEARE *406 515*

A windmill is marked, a little to the north of the known site, on Bowen's map of 1760, and at the remains of the tower a stone carved ' W.T. 1760 '[84] was formerly in existence, so unless this is the date of a re-building, Bowen's map would seem to suggest a former site. The Tithe Award shows several fields away from the mill tower called, for example, East Mill Furlong, perhaps the site of the earlier mill. The tower is parallel-sided and of lias rubble construction similar to its close neighbour, Ashton Mill, but it is built into an artificial circular mound some five feet high, similar to Stembridge Mill. The mill worked with a single pair of stones and a 4′ 6″ diameter French Burr has been used as a doorstep. Robert Hookway was miller in 1861, F. Hookway in 1865 and William Ham and Thomas Adams in about 1870. Henry Moon is recorded as miller in 1872 and 1875, and he is said to have been the last to work the mill, with John Peters. The mill is said to have ceased work in about 1880 because ' the horse of a man who was riding along the adjoining road bolted in fright at the sound of the whirling sails and threw his rider, killing him instantly. The windmill was never worked again '. Another local story concerned with one of the later millers who was sought by the police in connection with some offence, but who successfully evaded arrest because instead of being at home he had contrived a hiding place inside the windmill. In 1910-11 the mill tower and nearby miller's cottage were incorporated into a new dwelling house for a Miss Luttrell of Badgworth Court. This site is often referred to as being in Stone Allerton, a former part of Weare parish which was transferred to Chapel Allerton parish in 1885, but the remains of the windmill are within the present Weare parish.

WILLTOWN MILL, CURRY RIVEL, IN DECAY EARLY THIS CENTURY
The height of the tower can be gauged from the house in the foreground. The sails appear to be fixed to the windshaft on an iron cross.

Willtown Mill in about 1930, reduced to a stump which still remains to a height of about 9 feet.

Left: STEMBRIDGE MILL, HIGH HAM

A photograph by Henry Corder showing the mill with sails removed and the steam engine drive appearing at the back of the tower.

Below: STEMBRIDGE MILL AS RESTORED BY THE NATIONAL TRUST

The low building behind is the former bakery.

Right: 49″ diameter French Burr stones in Stembridge Mill, showing the patent balance weight pockets in the plaster back of the runner stone. The drive to the stones has gone.

Below: The windshaft and brake wheel in Stembridge Mill, from the tail. The compass arm notches can be seen between the clasp arms.

Left: The stone spindle bearing on the bridge tree in Stembridge Mill. The massive bridge post to the left has a twist peg in it, formerly used to control the angle of the shoe, and thus the supply of grain to the stones, by a string twisted round it.

Below: The mill trough of elm and the iron bevel gears which drove the second pair of stones in Stembridge Mill when steam power was used.

Right: 'Wind Mill new Built' at Week, Curry Rivel, from an estate map of 1768.

Below: Pibsbury Mill, Huish Episcopi, the notable exception in the form of Somerset tower mills. The photograph was taken early this century after the mill had ceased work for it had already lost its cap.

Above: HUTTON TOWER MILL IN ABOUT 1930

The tower has since been much rebuilt to a height of about 18 feet.

Right: KENN TOWER FROM THE SOUTH WEST

The carved stone over the ground floor window can be seen but since this photograph was taken in the 1930s the tower has lost its corrugated iron roof.

Above: The timber windshaft and clasp arm brakewheel in Vale Mill. The brake,of linked timber blocks, can be seen around the circumference of the brakewheel, with the brake lever to the bottom right.

Right: Mr Quick, the last miller at Vale Mill, standing before his mill in 1907. The mill has perhaps stopped work by wind at this time for there is no cloth furled on the sails.

VALE MILL

Left: The spur wheel at the foot of the upright shaft with the governor driven by belt from the shaft itself, in the foreground. The governor regulated both pairs of millstones, which were underdriven, and *above:* the secondary drive take-off from a crown wheel on the upright shaft over the stones. The top of the hopper over the stones can be seen, also the neat all-timber gearing and shafting.

MOORLINCH WINDMILL ON KNOLL HILL IN ABOUT 1840

A water-colour from the Braikenridge Collection, originally commissioned to illustrate Collinson's history of the county.

Above: An engraving of Portishead from the sea and showing the tower mill above the mill cottage to the right. The photograph of the remains of the tower illustrates the degree of artistic licence in this picture.

Left: The straight battered tower of Portishead Mill as it survives, looking north west over the roof of the mill cottage and Windmill Bay.

SHAPWICK

Above: The mound on which the post mill stood, a scene typical of many Somerset sites, and *below*, the base of the tower mill built into a mound, photographed in the 1930s. The remains are now much over-grown.

Uphill tower, with the church a shadow in the distance, on a wet day in the 1930s prior to the tower being rebuilt and castellated as an observatory.

WALTON MILL

Left: A photograph by Henry Corder taken early this century of the mill in early stages of decay. The pulley wheel mounted outside the door implies that a steam engine was used to drive the mill on windless days, and *right,* from the south in advanced decay in the early 1920s. The clasp arm timber brakewheel can be clearly seen.

Left: The tower of Weare Mill turned into an observation platform after being incorporated in a dwelling house in about 1911, and *right,* the tower of Weare Mill as it survives, with a flat roof replacing the platform. The high raised base around the foot of the tower can be seen.

A fine photograph of Heath House or Westfield Mill, probably sometime in the 1890s. One of the two men outside the mill door is William Tucker, the last windmiller here. This picture sums up very well the typical features of Somerset tower mills and the obvious pride of the men who worked them.

WEST MONKTON, BURLINCH

The very overgrown but complete shell of this tower mill survives in Burlinch Wood. The tower, of local shaly stone has opposite doorways at ground level and contained four floors.

WEST MONKTON, WOODBALL

The squat cylindrical tower forms part of a two storey house.

BROADFIELD MILL ON FELTON HILL, WINFORD

The tower of this former windmill is a prominent landmark opposite Bristol's Lulsgate Airport.

WOOLAVINGTON MILL, ABOUT 1910

A fine, elegant Somerset tower mill in the early stages of decay. The longitudinal bars which carried the sail cloth can be seen alongside the whips.

WORLE MILL

A painting of 1885 based on a sketch of 1871, which gives a good impression of the mill's lofty, exposed site.

WEDMORE

1. 441 474
2. 425 482
3. approx. 426 509

The parish of Wedmore once contained five windmills, the ' isle ' of higher ground on which the town stands giving ideal sites. A windmill is shown to the south east of the town on Bowen's map of 1760 and Day & Masters' of 1782, and stood at Maltfield Farm,[85] at the junction of Mill Lane and the road to Glastonbury via Westhay and Meare. It is said to have been a tower mill which worked with a single pair of stones and was ' built of very good stone '. In 1861 the miller was Edwin Harvey, the mill having been worked for much of its life by members of the Harvey family. After Harvey, in about 1875 the owner was a Mr William Millard, a corn merchant, who put an inexperienced man in charge and the mill was caught in full sail and in consequence was terribly smashed up. This happened in about 1877-8, and the tower was demolished in about 1886. The site has now been obscured by later building.

According to the *Wedmore Chronicle*,[86] Surgeon Westover made a will in 1678 and left his mill to his third son, Henry. This was the mill disestablished on 30 May 1700, by John Westover, brother of Henry, and the mill, or part of it, was removed to Mark. When the mill was willed to Henry in 1678 it apparently had not been grinding within living memory. It was described as being in the North field and the site can be determined from the Tithe Award, where ' Mill Batch ' and ' Westover Mill Tyning ' are given along Quab Lane, north west of Wedmore. A slight mound is still visible in the corner of the field.

' Burnt Mill ' and ' Burnt Mill Field ' appear at the third location right on the parish boundary with Chapel Allerton in the Wedmore Tithe Award. There is no visible evidence on site, but the crossing point of several footpaths coincide with one of the highest and most exposed sites on the Isle of Wedmore.

WEDMORE, Heath House *421 469*

Richard Emery is said to have been working a windmill at Heath House from 1737 to 1780, but the exact date of building of the last windmill to work at the hamlet of Heath House is unknown, although a beam in the mill is said to have been carved with initials, unrecorded, and the date 1775. There is also a local story of a windmill, milled by a man named Webber, being burnt out in about 1740, when the miller had fallen victim to Somerset ' Scrumpy '.

The last windmill to occupy this site was known as Heath House or Westfield Mill, presumably from having been built in the former west field of Wedmore manor. It was a stone tower mill of the typical Somerset parallel-sided form, with a thatched gable cap, four broad cloth-set sails and worked a single pair of 4′6″ French Burr millstones. The mill was last worked by William Tucker and two diaries kept by him for the years 1886 and 1887 give a little useful information about the mill and the miller's life.[87] In May 1886 William

Tucker bought reed for the mill from Mr Hole of Blackford and spent four days in June thatching the cap of the mill. Several entries relate to staying up until after midnight, or getting up in the early hours if the wind was right for milling, and William Tucker is said to have passed the time by playing his violin in the mill. E. Cooper came to dress the stones, from Cheddar, and John Latcham who ' repaired the wheel in mill ' in August 1887 was a carpenter in Heath House in the 1880s. The mill worked by wind until the end of the 19th century, a date variously given as 1892 and 1900, when it was damaged by a mid-summer whirlwind. William Tucker had a steam mill in a building adjoining his stable, across the yard from the windmill, which he used up to his death in 1902.

George Wilkins was miller at Heath House until 1810, followed by his son John until 1833. In the Registers of Wedmore church are the following interesting entries under Burials : ' George Wilkins of the mill an honest man, December 8th 1810 '. ' John Wilkins of Heath House (aged) 27 years, struck dead by lightning in the mill June 7th 1833 '. Edward Tucker was miller in 1861, Mrs Mary Tucker, perhaps his widow, in 1872 and John Spearing, who came from Watchfield Mill, is recorded as windmiller in 1875. From about 1876 to 1897 William Tucker is recorded as miller and corn-factor by wind-power, and in 1902 by steam-power. An enamelled sign on the gable of a building south of the mill house still reads ' William Tucker. Corn & Cake Merchant ' and it is said that he did a good business, keeping three men besides himself, also four horses and conveyances for collecting maize and barley from Cheddar and Highbridge stations and for delivery after grinding. He had several sheds put up and did a big trade in meal and corn for pigs and poultry. In 1902 William Duckett took over the mill and in 1909 an oil engine was bought for £180 and installed in place of the steam mill. This oil-engined mill ground about four sacks an hour when on full load, the whole output being some two hundred sacks of flour and six tons of cattle cake per week. Three men were employed, one to grind and two to fetch from the stations and deliver the flour in the villages with the aid of three horses. When at work by wind, Heath House Mill was said to have been in close competition with nearby Ashton Mill and, although disagreeing with any alleged supremacy by Heath House Mill, Tom Petheram recalled 'Mr Stevens' two carts and Mr Tucker's two carts used to load at Cheddar station from railway trucks and then travel together through Cheddar Moor to Cocklake, then we turned up to Rughill. Tucker's men turned left and on through Wedmore to Heath House and we travelled up Rughill, came up the hill to Crickham and Stoughton Cross and then turned right to Washbrook and Ashton Mill. Mr Stevens and Mr Tucker at times took maize and barley from the station to Cheddar Mill to be ground and then loaded home with meal '. All Mr Stevens would comment was ' Tucker was a very lucky man. He had three babies within eleven months of being married, one on the night of his wedding and twins within the eleven months after. Lucky? Sure ! '.

The tower of Heath House Mill, without sails or cap, stood in the mill yard until 1962 when it was pulled down by tractor and much of the stone ' turned to dust '. During the 1920s the yard had been used by Mr Tucker's son, Stanley, for a wheelwright's business, and in the 1930s the tower was used as a fowl-house, a use which discouraged close investigation. A new house has been built on the site of the tower and remains of the two mill-stones can be seen in the front garden. The modest double-fronted mill-house, rebuilt in 1856 after a fire, stands a little to the south.

WEDMORE, West Stoughton *419 487*

A windmill is marked south of the cross roads to the south west of West Stoughton on the 1st edition OS map of 1817. It is said to have been a stone tower mill, known as Stoughton Mill, which worked with a single pair of stones until about 1875. The last miller was Mr John Burrows. A mound, with no masonry whatever visible, can still be found.

WELLINGTON

At Burgage, in the centre of Wellington, a circular building which no longer exists, was said to be ' the trunk of a well-known old windmill '.[88] The old name of North Street, just outside Burgage, was Millway, but the history of this windmill is obscure. It is also said to have been used for grinding oak bark for the tanning industry and it is therefore just possible that the circular building may have been built or used for a horse-mill, rather than a windmill.

WESTBURY *approx. 497 482*

South west of Westbury-sub-Mendip a conical hill is known as Windmill Hill.

WEST HUNTSPILL *1. approx. 310 453*
2. approx. 342 451

On John Ogilby's road map of 1675 a windmill is shown close to the main road at Huntspill, a site marked as ' Millbatch ' in the Tithe Award. A Mill Lane exists at the second site and ' Millbatch Orchard ' appears in the Tithe Award.

WEST MONKTON, Burlinch *265 295*

In 1364 there were three mills recorded in West Monkton, one of which was Bathpool watermill, when the manor was part of the estates of Glastonbury Abbey, and at the time of the Dissolution it appears that among the King's acquisitions were 'A mill demised to Roger Adamps at a yearly rent of 23s.4d ' and 'Another mill demised to William Adamps at the same rent '.[89] These are thought to have been windmills, and the remains of two exist in the parish. That at Burlinch is marked on the 1st edition OS map, 1817, but the Common on which it stood was enclosed, and subsequently became depopulated, in 1810, and perhaps it was disused from about that time, although a local villager who died in 1952, aged 80, could remember corn being brought to the mill by pack-horse as a child. This implies a last working date of about 1880, but there are no records of millers here in the 19th century. The remains consist of a parallel-sided tower now heavily overgrown, and the tower is completely gutted. At what was second floor level a fireplace exists in the tower wall and internally the two lower floors were originally plastered. This tower, like that at Brockley, is difficult to locate and even more difficult to photograph.

WEST MONKTON, Woodball *255 293*

Although not marked as a windmill on any of the county maps, this tower is perhaps the descendant of one of the two unspecified mills in West Monkton in the late middle ages. The tower is now part of a house, less incongruously connected than other such conversions, and magnificently sited above the 400 foot contour with nothing of comparable height around it and with reasonable access. The tower wall is 3′ 6″ thick and a window at present first floor level has a recess below it, in which it is possible to stand or sit, similar to the second floor window of Ashton Mill.

WEST PENNARD *approx. 560 382*

The tolls of Pennard windmill, 1 quarter 6 lbs. of wheat and beans, are recorded in the Granger of Glastonbury Abbey, 1361-2.[90] 'Windmill Field' is recorded in the Tithe Award.

WESTONZOYLAND *1. 339 350*
2. approx. 365 346

In the 1330s several mills formed part of the estates of Glastonbury Abbey in this area, including a windmill at New Westonzoyland valued at £2.13s.4d.[91] Greenwood's map of 1822 shows windmills at Penzoy and Westonfield; the former site is also marked on the early OS maps. The second site is now obscured by an airfield, built during the last war, and the miller of the windmill which stood there was a Mr Laver in 1833.

WILLITON *approx. 102 432*

'Mill Meadow' and 'Great Mill' occur in the Tithe Award at this coastal site a little to the west of the two possible coastal sites at East Quantoxhead. At West Quantoxhead, south of the church, is a public house called 'The Windmill' (103 418) with a steel wind-pump close by it.

WINCANTON *approx. 719 290*

A Windmill Hill exists and a Windmill Farm has been so-called from at least the 17th century at which date there was no specific reference to an extant windmill.[92]

WINFORD, Felton *515 649*

The site of Broadfield Windmill is one of the few north of the Mendip Hills and is very exposed to the prevailing south-westerly weather, being some 600 feet above sea level. The parallel-sided stone tower of the mill survives as a dwelling, but sadly little of its history is known. The windmill probably ceased work in the late 1880s and was converted into a house soon after by the addition of a small lean-to extension and a polygonal tiled roof. The tower now contains three floors, the lower of which is about 6 feet below the present ground level and was used as the wash-house by the present resident's father who worked at the nearby red ochre works. Just outside the entrance to the tower the remains of one or two broken sandstone conglomerate millstones have been built into the floor.

If this is the windmill Hay Wilson wrote about in *A Somerset Sketch Book*, published in 1912, it would appear that the mill was not remembered as working, for in writing about the children playing on 'Wansdyke Common' he says 'none of their fathers can remember seeing sails on the disused windmill'. He continues 'Since the windmill was reduced to the footing of a mere living house, the dwellers have planted a screen of gaunt thorn-trees on the seaward side because the fiercest wind blows from that direction, and all the trees lean away from it, so that the place looks storm-blown even when the breeze is slight, although that happens seldom'.

WOOLAVINGTON *347 408*

Two windmills were recorded at Woolavington in 1222[93] and a windmill in 1630 was rented and occupied by William Burge of Bawdrip and Lawrence Clappe of Woolavington, who charged one Robert Bennet, a labourer of Cossington, with stealing grain from their mill. Bennet was examined before a Justice of the Peace, Abraham Burrell by name, on 10 February 1630.[94] A windmill is marked here on the maps of Strachey, 1736, Bowen, 1760, Day & Masters, 1782, and Greenwood, 1822. A drawing of the tower mill appears on a Map of the Manor of Woolavington Pym of c.1775 when the lessee of the mill cottage was George Luke. G. Luke is recorded as miller in 1861, Samuel Luke from about 1866 to 1875, Denis Denman came to Woolavington from Middlezoy windmill about 1880 and left for Moorlinch windmill in about 1889. The last miller was George Cox, who took over from Denman and stayed until his death in 1906 when he was not replaced. The mill is said to have stopped work about 1904.

Woolavington windmill was a tower mill of the usual Somerset pattern, the tower wall being nearly parallel, twelve feet diameter at the base and eleven feet two inches diameter at the curb. The cap, which survived until about 1930, was thatched, the winding gear contained in a tail-box and the cap turned on a dead curb, riding on blocks of wood about four inches square. The windshaft, of oak, carried a compass arm brakewheel of elm, which was approximately seven feet in diameter with four arms ten inches deep by four inches thick. The sail stocks were carried in a cast iron poll, said to be dated 1830, and the mill drove a single pair of 4′ 3″ diameter stones, direct from the brakewheel. The tower contained four floors and, after the mill was disused, stood in fair preservation for many years, losing its cap in a gale in about 1930, but it was roofed over, in thatch, for the owner, Miss H. M. Collings, who loved the mill at a time when such care and expense was a rare happening in Somerset. The tower was finally demolished in about 1967, the only survival being the tail end of the windshaft which remains, as a post for climbing roses, in the garden of the original mill house.

WORLEBURY *approx. 326 627*

Strachey's map of 1736 and Bowen's of 1760 show a windmill towards the seaward end of Worlebury Hill, roughly on the site of the present reservoir. The site would have been well exposed, but reasonably accessible from Kewstoke or Weston. It is now thickly wooded.

WORLE *352 632*

A windmill is recorded at this site on Bowen's map of 1760 and Day & Masters' of 1782 on this striking position on the eastern shoulder of Worlebury Hill, overlooking a great tract of sea and land. In 1829 the village of Worle was described as ' sheltered from the sea wind, by its remarkable hill, crowned with a wind-mill '[95] and in 1849 Matthew Day was residing in furnished lodgings at Prospect House, Worle ' and during such residence carrying on business as a Miller, at Worle Mill, Worle aforesaid, and also carrying on business as a Salt Merchant, Flour Factor and General Dealer at Weston Super Mare '.[96] In 1861 William Rogers was miller and shopkeeper and in 1870 he advertised the mill to be let or sold as a well-established business and ' The mill contains one pair of stones, two dressing and smut machines, is in complete repair, has a bakery business in connection with it, and is only given up by the proprietor on account of ill-health, the mill having been in his possession for 30 years '.[97] By 1883 the mill was marked on the OS maps as ' Worle Tower ', so William Rogers had presumably failed to find a purchaser. A local guide book of 1885 refers to the 'old and gradually decaying structure ' and in about 1889 plans were drawn up by the son of a local doctor and the tower was raised in height and ' built around with extensions and ornamentations ' apparently as a private observatory. 'Afterwards, it was used as an observation tower conducted on a commercial basis but the venture proved unsuccessful ' and the property was bought by a local solicitor and dubbed, though not by him, ' Baker's Folly '.[98] A bungalow was later erected nearby and the whole occupied as a private residence under the style of ' The Observatory ', in which guise it has passed through many ownerships to the present. Wheat was grown on Worlebury Hill, as at Uphill and Brean Down, probably during the Napoleonic Wars, and the hill also had a good reputation for mushrooms, as well as being a traditional site for Bank Holiday games and picnics with the inhabitants of Worle and Milton. The heightened and castellated tower of the former Worle Mill still occupies its prominent position above Worle village.

YEOVIL *approx. 565 162*

A Windmill Hill exists north of the railway and River Yeo, but has been renamed within recent years.

YEOVIL WITHOUT *approx. 533 182*

A site between Yeovil and Chilthorne Dormer is suggested by a field name in the Tithe Award.

NOTES AND REFERENCES

Abbreviations used in the notes:

BCA	Bristol City Archives
SA&NHS	*Somerset Archaeological & Natural History Society Proceedings*
SDNQ	*Notes and Queries for Somerset and Dorset*
SRO	Somerset County Record Office, Taunton
SRS	Somerset Record Society publications
VCH	*Victoria County History, Somerset*
WM	*Weston Mercury & Somerset Herald*
WRO	Wiltshire County Record Office, Trowbridge

The following sources have been generally used:

MAPS:

Map of Somersetshire, John Speed, 1610.
A Map of Sedgemoor, with Adjacent Places, Ricardus Newcourt, c.1660. Copy in SRO.
Britannia, John Ogilby, c.1675. Maps 11, 32, 58 and 60 refer to roads in Somerset.
Somerset Surveyed and Protracted, John Strachey, 1736.
Map of Somerset, Emanuel Bowen, c.1760.
A Map of the Country 11 Miles around Bristol, Benjamin Donn, 1769.
A Topographical Map of the County of Somerset, W. Day & Thomas Masters, 1782.
First Edition Ordnance Survey (OS), 1″ to 1 mile, 1805-44. Somerset published 1809-17.
Map of the County of Somerset, surveyed 1820-1, C. & I. Greenwood, 1822.
Large Scale Ordnance Survey, 1882-8, published 1884-90. (25″ and 6″)
Tithe Awards and Maps of 1830s-40s in SRO.

TRADE DIRECTORIES:

Bragg, 1840.
Harrison, Harrod & Co., 1859.
Kelly, 1861, 1875, 1883, 1889, 1894, 1897, 1902, 1906, 1910, 1914.
Morris & Co., 1872.
Pigot, 1842.

ARTICLES PREVIOUSLY PUBLISHED FOR THE AUTHORS:

'Windmills in Somerset', A. J. Coulthard, *SDNQ* Vol.26 1955. Part 1, 84-7; Part 2, 107-8; Part 3, 139-40; Part 4, 161-3.

'Ashton Mill', A. J. Coulthard, *SDNQ* Vol.26, 1955, 199. Vol.27, 1961, 10-11, 76-8.

'Recollections of Some Somerset Millers', A. J. Coulthard, *Millnotes* 3, 1972, 34-7.

'Heath House Windmill, Wedmore, Somerset', Martin Watts, *Millnotes* 5, 1973, 84-6.

'Windmills of Somerset', Martin Watts, *Bristol Industrial Archaeological Society Journal* 6, 1973, 21-31.

Somerset Windmills, Martin Watts, Keynsham, 1975. (Revised off-print of *Bias Journal* 6 article).

'Documentary Evidence of Windmills at Walton, Somerset', Martin Watts, *Somerset Industrial Archaeological Society Journal* 2, 1976.

TEXTUAL NOTES

Chapter 2. HISTORICAL BACKGROUND

1. See, for example, 'The Granger of Glastonbury Abbey, 1361-2' and 'Mills on the Estates of Glastonbury Abbey in the Later Middle Ages', Ian Keil, *SDNQ* 28, 1968.
2. *VCH* 3, 1974.
3. *SRS* 8, 1894. We are indebted to Dr Dunning for this reference and dating.
4. *SRS* 57, 'Collectanea III', 1942.
5. *SDNQ* 26, 1955, 159.
6. *Transactions of the Newcomen Society*, 34, 1961-2.

Chapter 3. DISTRIBUTION AND DESIGN

1. *General View of the Agriculture of the County of Somerset,* John Billingsley, 1797.
2. *Somerset Countryman* 2, No.111, July 1932, 36. 'On Somerset Windmills', Henry Corder.
3. SRO.
4. See *Bias Journal* 6 and *Somerset Windmills,* Martin Watts, for a contemporary survey of remains in the county.
5. *WM*, 5 March 1870.
6. *The Draining of the Somerset Levels*, Michael Williams, 1970.

Chapter 5. SOMERSET WINDMILLS AND SITES

1. *VCH* 3, 1974, 68.

2. WRO.
3. *SDNQ* 28, 1968, 181-4.
4. Copy of c.1830, Church Commissioners Collection, SRO.
5. *The Draining of the Somerset Levels,* Michael Williams, 1970, 94-5.
6. *SDNQ,* op cit.
7. *The History and Antiquities of the County of Somerset,* 2, John Collinson, 1791, 121.
8. Sessions Rolls, 64 Pt.2. Fol.2231, SRO.
9. *Delineations of the North Western Division of the County of Somerset,* John Rutter, 1829, 31.
10. Sun Fire Insurance Policy No. 514003. Simmons Collection.
11. *WM,* 3 July 1869.
12. Information from Trevor Miles, archaeologist.
13. Royal Exchange Fire Insurance Policy No. 199289. Simmons Collection.
14. *SDNQ* 29, 1974, Pt.292.
15. *Bristol Spelaeological Society Proceedings,* 8, No.2 1957-8, 89-96. 'Barrow and Windmill at Butcombe, North Somerset', P. A. & M. H. Rahtz.
16. *Axbridge Caving Group & Archaeological Society Journal,* 2, No.4, 1955.
17. *SA&NHS Proc.* 45, 1899. 'Descent of the Manor of Allerton', Prebendary Coleman 25-50.
18. *SA&NHS Proc.* 46, 1900. 'The Manor of Allerton and its Tenants', Prebendary Coleman. 64-108.
19. Estate Map of 1797 by William White, SRO.
20. A beam above the stone floor bears the initials, name and date 'Paul M. Wilkins, June 2nd 1799'. The Paul Wilkins who worked at Ashton Mill until about 1866 would have been ten years old at that time.
21. *The English Windmill,* Rex Wailes, 1954, 69-70.
22. *SA&NHS Proc.* 46, 1900.
23. Information from the present incumbent, courtesy A. G. Shove.
24. *VCH* 3, 1974, 106-7.
25. *Yorkshire Gazette,* 15 December 1827 (Vide Taunton Courier). Simmons Collection.
26. *SDNQ* 12, 1911, 342.
27. *The Heart of Mendip,* F. A. Knight, 1915.
28. SRO.
29. *Sherborne Mercury,* 30 January 1775. Reference kindly supplied by H. L. Douch.
30. SRO.
31. *SA&NHS Proc.* 113, 1969 and 115, 1971. 'Somerset Barrows' Parts 1 and 2, L. V. Grinsell.
32. *SA&NHS Proc.* 113, op cit.
33. *SA&NHS Proc.* 115, op cit.
34. *SDNQ* 28, 1968, 181-4.

35. Collinson, op cit, 268.
36. *Somerset County Herald,* 31 July 1926.
37. *Antiquities,* Vol.III, F. Grose, 1775.
38. Williams, op cit, 112.
39. *The Diary of a West Country Physician, AD 1684-1726,* Ed. E. Hobhouse, 1934. Entry for 5 November 1723.
40. Estate map, surveyed 1742 drawn 1767, SRO.
41. Map of the Manor of Netherham, 1779, SRO.
42. Ibid.
43. *Stembridge Tower Mill, High Ham, Somerset,* Kathleen D. Taylor, N.D.
44. *SDNQ* 29, 1974, 173.
45. *VCH* 3, 1974, 8.
46. *Somerset County Herald,* undated cutting. (Late 1930s.)
47. *In the Smoke of War,* Walter Raymond, 1895.
48. *VCH* 3, 1974, 7.
49. We are indebted to Mr H. Coward of Hutton for notes from his research on the history of the parish.
50. Estate Records of Smythe Family, BCA.
51. *Weston-super-Mare Gazette,* 15 May 1854.
52. Information from H. Coward.
53. The initials were said to stand for 'C. Brakenbridge' or perhaps 'Brakenridge' but this is as yet unsubstantiated.
54. *VCH* 3, 1974, 117.
55. *VCH* 3, 1974, 26-7, and verbal information from Dr R. W. Dunning.
56. Collinson, op cit, 290.
57. *VCH* 3, 1974, 162.
58. Estate Map, 1758, SRO.
59. *Wedmore Chronicle* 2, 1887-1898, 140.
60. *VCH* 3, 1974, 220.
61. An extraordinary account of Moorlinch windmill in the 17th century can be found in *Second Time Round,* E. W. Ryall, Jersey, 1974, in which the author recalls a former life. The mill is described as being a sunk post mill.
62. Braikenridge Collection, Taunton Museum.
63. Survey of the Manor of Latterford, 1766, SRO.
64. *SA&NHS Proc.* 115, 1971, 110.
65. Correspondence with Somerset River Authority.
66. *VCH* 3, 1974, 54.
67. BCA. (Papers 03280.1-.5).
68. *Somerset Memories and Traditions,* Frances H. Wood, 1924.
69. Watercolour in possession of Miss Wilmott in 1930s.
70. *SRS* 8, No.93, 1894, 150.
71. *SDNQ* 28, 1968, 181-4.
72. *Somerset I.A. Society Journal* 1, 1975, 16-17. 'Shapwick Windmill', C. A. Buchanan.

73. *VCH* 3, 1974, 144-5.
74. Collinson, Vol.1, op cit. 248.
75. Estate Map, 1787, SRO.
76. *North Curry: Ancient Manor and Hundred,* Hugh P. Olivey, 1901, 230.
77. *SDNQ* 28, 1968, 181-4.
78. *Seaboard of Mendip,* F. A. Knight, 1902, 265.
79. Simmons Collection.
80. Rutter, op cit, 77.
81. *Transactions of the Newcomen Society,* 34, 1961-2, 151-4.
82. Longleat Papers at WRO. See 'Documentary Evidence of Windmills at Walton, Somerset', Martin Watts, SIAS Journal 2, 1976.
83. Wood, op cit.
84. The initials 'W.T.' were said to stand for 'William Thomas', but this has not been substantiated.
85. Information from Mr F. J. Pearce, and see his *A History of Wedmore,* Somerset Archaeological Society, September 1971.
86. *Wedmore Chronicle,* op cit.
87. Courtesy of Miss Ruth Tucker.
88. *Somerset County Herald,* 16 September 1933, and information from A. G. Shove.
89. *SA&NHS Proc.* 90, 1944, 39-61. 'Notes on the Manor and Parish of West Monkton', E. E. Trotman.
90. *SDNQ* 28, 1968, 86-90.
91. *SDNQ* 28, 1968, 181-4.
92. *A History of Wincanton, Somerset,* George Sweetman, 1903.
93. SRS 57, 1942. 'Collectanea III'. 26.
94. Sessions Rolls 64 Pt.2, Fol.1371. SRO.
95. Rutter, op cit, 39.
96. *London Gazette,* 16 March 1849. Simmons Collection.
97. *WM,* 5 March 1870.
98. *WM,* 14 July 1934, 27 July 1940, 3 August 1940, 10 August 1940, 21 September 1940, 3 April 1970.

APPENDIX I

SOME MILLING FAMILIES OF SOMERSET

The accompanying charts show the close relationship between some of the milling families. Most of this information was collected in general conversation and is therefore limited to recollections and hearsay, and research would probably considerably extend it and fill in some of the gaps.

EMERY

Members of the Emery family were working in several Somerset mills, mainly watermills, in the 18th century and their names and respective mills are given with the date when they were known to have been at work:

Isaac	Wrington Mill	1700	
Thomas	Banwell Mill	1711	
John	Banwell Mill	1750	
George	Banwell Mill	1810	
John	Portishead Mill	1730	(not the windmill)
Thomas	Wraxall Mill	?	
Nathaniel & William	Compton Bishop Mill	1730	

Also members of the family at Glastonbury Mill (water ?) 1700, Portbury Mill and Kingsbury Mill in 1750. From 1737-1780 Richard Emery worked Heath House Windmill.

Pedigree charts of the Spearing, Stevens-Loader, and Wilkins families appear on the following two pages.

SPEARING

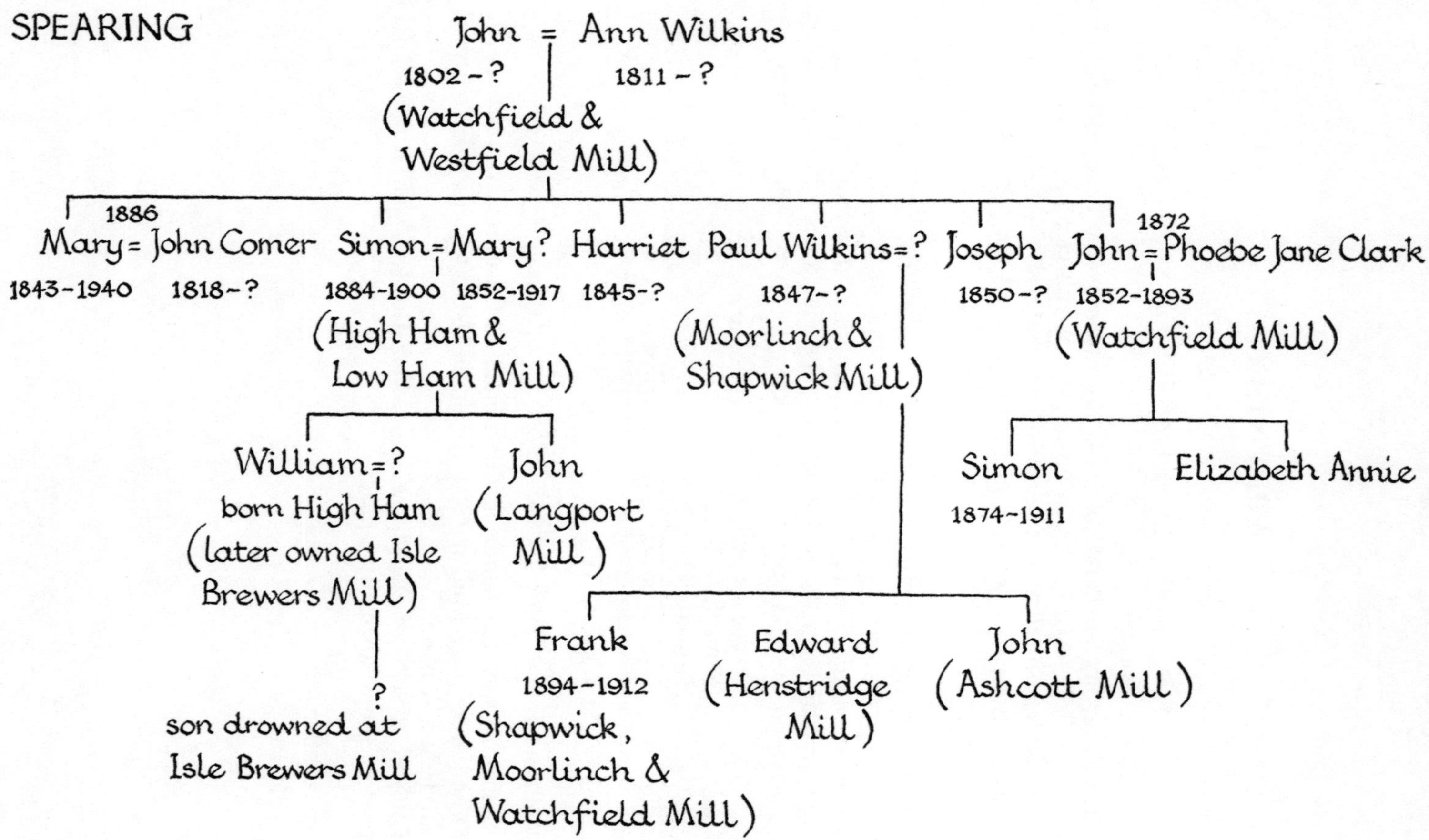

John = Ann Wilkins
1802 – ?
1811 – ?
(Watchfield & Westfield Mill)
1886
Mary = John Comer
1843–1940
1818–?
Simon = Mary?
1884–1900
1852–1917
(High Ham & Low Ham Mill)
Harriet
1845–?
Paul Wilkins = ?
1847–?
(Moorlinch & Shapwick Mill)
Joseph
1850–?
1872
John = Phoebe Jane Clark
1852–1893
(Watchfield Mill)
William = ?
born High Ham
(later owned Isle Brewers Mill)
John
(Langport Mill)
?
son drowned at Isle Brewers Mill
Frank
1894–1912
(Shapwick, Moorlinch & Watchfield Mill)
Edward
(Henstridge Mill)
John
(Ashcott Mill)
Simon
1874–1911
Elizabeth Annie

STEVENS-LOADER

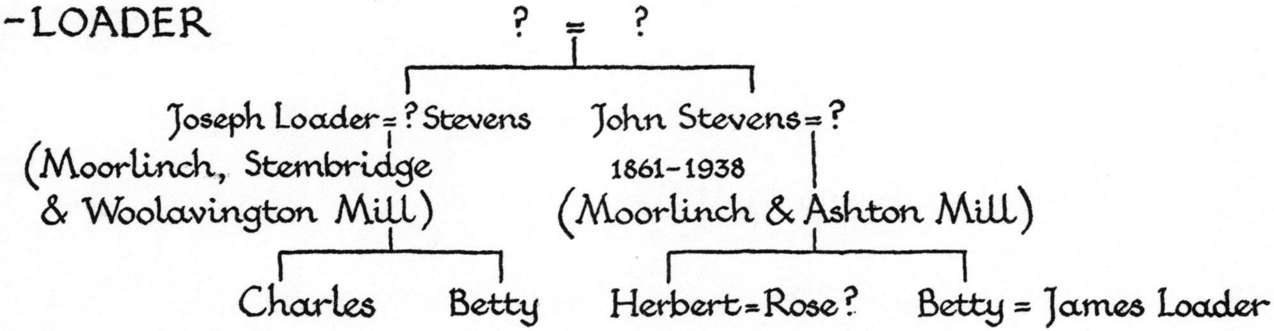

WILKINS

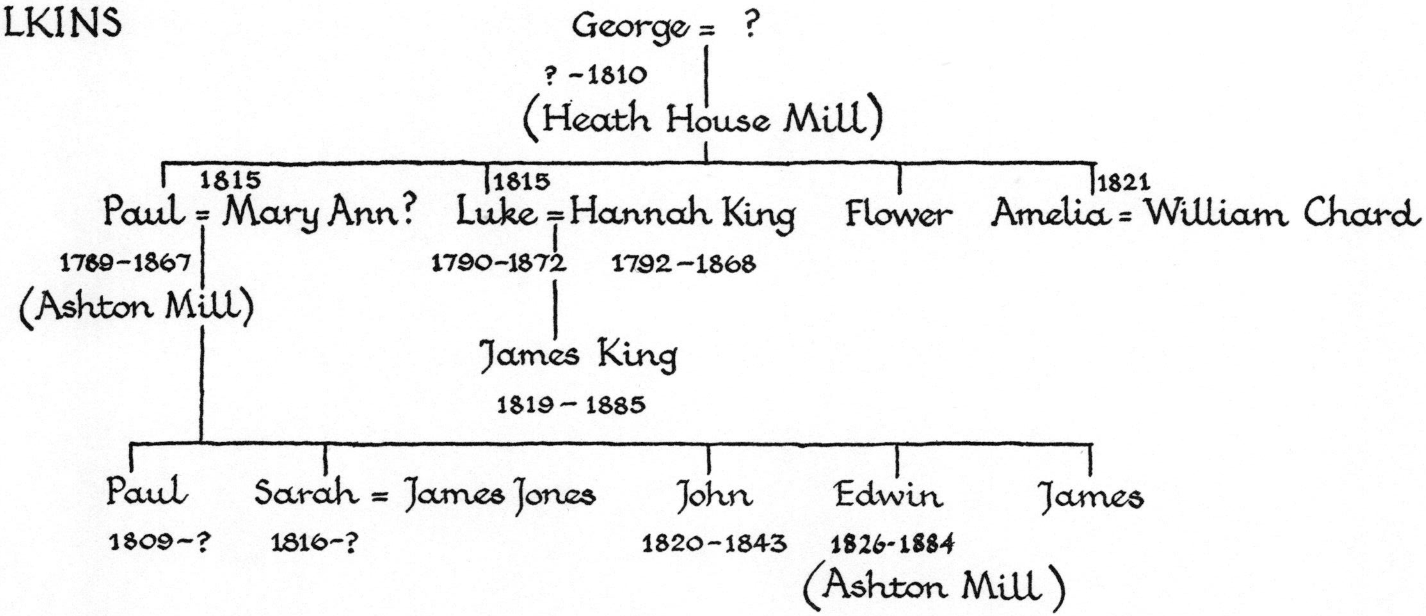

APPENDIX II

WIND ENGINE AGREEMENT

The following agreement, made in 1700 between a millwright on one part and a gentleman, a yeoman and a coal-miner on the other, is for the erection and maintenance of a wind engine to draw water by pumps from a depth of twelve fathoms at a coal working in the South Park at Wraxall, in north Somerset. The agreement is among records recently deposited by Sir Henry Calley at the Wiltshire Record Office (WRO 1178) and is transcribed in full by courtesy of the County Archivist. Particular thanks are due to Andrew Crookston and Kenneth Rogers at the Record Office, for bringing the document to notice and helping with its transcription.

This agreement serves to emphasise the importance of motive power in a growing industrial context; water-driven pumps were used at Somerset coal workings for drainage from the early 17th century, and steam power was introduced in the middle of the 18th century. The coal work in the South Park at Wraxall was part of the Nailsea coalfield, an area which evidently yielded good fuel, but the amount of water and dead ground met with in relation to the coal extracted made it an uneconomical area to work in later years.

Articles of agreement had made and agreed on by and between John Bryant of Portbury in the County of Somersett Millwright and William Wynter of Wraxall Gent John Godwin of Nalsea yeoman and Thomas Mudge of Nailsea aforesd coleminer All in the sd County of Somerset this 30th day December 1700 As followeth that is to say

Imprimis the sd John Bryant doth by Gods permission undertake to Erect and finish an Engine to go with ye winde in manner as a windmill in ye Parks called the South Parke in Wraxall to draw water out of the Pitt now begunne there with the help of Pumps from the depth of Twelve fathom That shall carry off a Shoote of water to ffil a Bore of Six Inches Diameter which sd Engine shalbee reddy to worke and carry off the whole shoot of water belonging to the coaleworks as the sd coaleworks shall from time to time bee wrought by the 28th day of February next

Item The sd John Bryant is to finde all materialls of Timber Lime and Ironworks And all other materialls of Pumps and other necessaryes for the perfecting the sd Engine And to bring it all in place Except the timber which hee buyeth of Edward Gorges Esquire which the sd Edward Gorges Esquire is to bring in place at his owne charge And the sd John Bryant after the sd Engine is finished is to keep the same in all needfull repayracon dureing the Term of Seaven yeares if hee the sd John Bryant shall soe longe live They the sd Wm Wynter John Godwin and Thomas Mudge doing or causeing to bee done noe willfull damage to it and the sd John at the end of ye sd Tearm of Seaven yeares will yeild upp ye sd Engine unto Edward Gorges Esq or his Assigns in good repayre as before mencioned and intended

Item They the sd William Wynter John Godwin and Thomas Mudge In consideraton of ye premisses hath paid in hand to the sd John Bryant Twenty one shillings and Six pence and will pay ffiveteene pounds more to him the sd John Bryant as hee shall have accausion to make use of it for buyeing Timber and materialls for ye same and ye sume of Twenty and ffive pounds more when the Engine soe undertaken as aforesd shalbee compleately finished to doe the worke according to the promise of him the sd John Bryant in full of fforty one pounde one shilling & six pence And likewise they the sd William Wynter John Godwin and Thomas Mudge will alsoe pay unto the sd John Bryant ffortye shillings per an by foure Equall quarterly payments during ye above sd Tearm of Seaven yeares in consideracon whereof hee the sd John Bryant doth hereby undertake and promise to keep in repayre the sd Engine if hee the sd John Bryant live soe longe

Signed Sealed and Delivered the paper
being first stampt according to
Law in ye presence of
Robert Baker
William Godwin
Lewis Donne

Sig
John (mark) Bryant

GLOSSARY

Bedstone	The lower, fixed, stone of a pair of millstones.
Bin	Storage compartment for loose grain, not usual in Somerset windmills.
Bolter	A machine for separating flour from bran by power-assisted sifting through a bolting cloth.
Brake Wheel	The primary gear wheel mounted on the windshaft on the rim of which the band brake acts.
Cap or Head	The moveable top of a tower mill which turns on a curb on the top of the masonry tower to bring the sails into the wind. In Somerset the term 'Head' was used by the millers, who spoke of 'bringing her head into the wind'. In relation to the short, squat tower of the Somerset mills it formed more of the structure than the word cap would suggest.
Clasp Arm	The construction of a wooden gear wheel in which the arms which support the rims are boxed around the shaft, clasping it and each other.
Cogs	Wooden teeth driven singly into mortices in wood or iron gear wheels.
Common Sails	The traditional cloth-set sail. Also known as 'Cloth sails' and, in Somerset, 'Plain sails'.
Compass Arm	The construction of a wooden gear wheel in which the arms are morticed through a wooden shaft.
Cross	A casting mounted on the end of a windshaft to carry the sails on its arms. In Somerset the known crosses were four-armed.
Crosstrees	Main horizontal beams in the sub-structure of a post mill.
Crown Wheel	A face gear mounted on the upright shaft from which secondary drives, to a sack hoist or dressing machine, for example, were taken.
Curb	The track on the top of the masonry tower of a mill on which the cap or head turns. Ashton Mill and Stembridge Mill have shot curbs, that is, live curbs with iron rollers, and Woolavington Mill had a dead curb, the cap being turned on wooden blocks.

FRENCH BURRS — French stones or burr stones. Millstones built up by binding together shaped blocks of stone imported from the Paris basin, the best stones for wheat milling.

GOVERNOR — Automatic device to maintain a constant gap between the millstones when milling.

GUDGEON — An iron journal mounted in the end of a timber shaft.

HOPPER — Grain container mounted over the millstones.

HORSE — Framework on top of the stone casing supporting the hopper and shoe.

LANTERN GEAR — An early form of gearwheel in which a circle of staves held between two wooden flanges served the purpose of cogs. It was usually associated with a driving wheel fitted with peg-like cogs which meshed with the staves of the lantern gear.

LEADING BOARDS — Longitudinal boards fixed along the leading edges of sails. On some Somerset mills cloth was set on the leading edge, while on others broad leading boards were fixed, for example, Watchfield tower mill.

LUFFING — cf WINDING.

MILL MOUND — Generally any raised ground upon which a windmill was constructed. In Somerset both post and tower mills were erected on mounds, sometimes because a tower succeeded a post mill and the latter were frequently built on mounds to give them extra height and sometimes with the sub-structure in mounds, to give extra stability. Some tower mills have an earth mound with a walled perpendicular edging, as at Stembridge Mill and Weare Mill, presumably to keep animals away from the whirling sails and to provide a working platform for the miller when setting or taking in sail and winding the mill.

NECK JOURNAL — The front bearing of a windshaft, turning in a neck bearing. The journal is of iron, the bearing, in Somerset, often of stone.

OVERDRIFT — Millstones driven from above, syn. overdriven.

PINION — The lesser of two wheels in gear.

PLAIN SAILS — cf COMMON SAILS.

POLL END — Cast iron fitting on the end of the windshaft into which the sail stocks are fixed. The fitting consists of two iron boxes at right angles to each other; formerly the sail stocks were morticed through the end of the timber windshaft.

POST MILL — A windmill of which the timber-framed body, carrying the machinery and the sails, turns on a vertical post.

QUANT — The spindle carrying the stone nut which drives an overdrift stone.

QUARTERBARS	The diagonal timbers in the sub-structure of a post mill which brace the post from the crosstrees.
ROLLER MILL	A form of corn mill developed in the 19th century in which steel rollers replaced millstones to grind the corn.
RUNNER STONE	The upper, driven stone of a pair of millstones. EDGE RUNNER STONES are vertical stones which revolve on a circular base and were used for grinding such materials as oak bark for the tanning industry.
SACK HOIST	Wind-powered mechanism for raising sacks through the mill.
SHEERS	The two main timbers extending from breast to tail in the cap of a mill. Also the two timbers below a post mill body.
SMUT MACHINE	A type of dressing machine to remove smut from grain.
SPUR WHEEL	Also Great Spur Wheel. The gear mounted on an upright shaft which drives the stone nuts.
STOCKS	The timbers to which the sails are fixed. A four-sailed mill has two stocks, each of which is passed half way through the poll end and wedged in place, and each then carries two sails.
STONE DRESSING	The craft or skill of sharpening the millstones prior to grinding.
STONE NUT	The final driven pinion in the drive to the stones. In an overdrift mill this gear is mounted on the quant, in an underdrift mill on the stone spindle.
STONE SPINDLE	An iron spindle on the top of which the runner stone is carried.
SUNK POST MILL	A post mill of which the sub-structure is buried in the ground or in an artificial mound.
TAIL BEARING	The journal bearing at the rear of a windshaft.
TAIL BOX	The box under the rear of the cap which contains the winding gear. A typical feature on Somerset tower mills.
TAILPOLE	A timber lever for winding the mill body, in the case of a post mill, or cap in the case of a tower mill, manually or with the aid of a winch. At Kenn Mill a pony is said to have been used to assist in winding the mill.
TOWER MILL	Masonry towers of which only the cap or head moves to bring the sails into the wind.
UNDERDRIFT	Millstones driven from below, syn. underdriven.
UPRIGHT SHAFT or	VERTICAL SHAFT. The main shaft driven by the wallower and carrying the spur wheel for transferring drive to the stones and a crown wheel for taking off secondary drives.

INDEX

WINDMILLS AND SITES

MILLERS AND MEN